Shorebirds of North America

Shorebirds of North America

The Photographic Guide

Dennis Paulson

Princeton University Press
Princeton and Oxford

Published by Princeton University Press, 41 William Street,
Princeton, New Jersey 08540

Library of Congress Cataloging-in-Publication Data

Paulson, Dennis R.
Shorebirds of North America : the photographic guide / Dennis Paulson.
p. cm.
Includes bibliographical references (p.) and index.
ISBN 0-691-10274-0 (cl : alk. paper) — ISBN 0-691-12107-9 (pbk. : alk. paper)
1. Shore birds—North America—Identification. 2. Shore birds—North America—
Pictorial works. I. Title.

QL681.P37 2005
598.3′3′097—dc22 2004044645

British Library Cataloging-in-Publicaton Data is available

This book has been composed in Stone Serif Family & Gill Sans Family

Printed on acid-free paper. ∞

www.nathist.princeton.edu

Composition by Bytheway Publishing Services

Printed in Italy by Eurographica

1 3 5 7 9 10 8 6 4 2

Contents

Preface

Shorebirds, or waders, are among the most engaging birds in the world. Diverse enough to present identification challenges and to satisfy listing urges in the most hardcore birder, they are interesting enough in their ecology and behavior to present other kinds of challenges to the dedicated ornithologist and spectacular enough in their flocking and migration to satisfy the casual observer for whom nature represents pure aesthetic enjoyment.

Coastlines and freshwater wetlands are exciting places to visit, and their complement of shorebirds, changing over the day with the tides and over the year with the birds' migration cycles, makes them even more attractive. A shorebird enthusiast gets the benefit of scenery on all sides, because most shorebirds are birds of the open, whether a windswept salt marsh, a wave-crashed jetty, a mangrove-lined estuary, an alkaline playa, or a tussocky tundra.

Few birds are such far-flung migrants as shorebirds, so their seasonal movements, with sudden appearances and disappearances, increase the level of interest in the group. During spring and fall migration, there are so many species and individuals coming and going that one literally should place oneself in shorebird habitat every day to intersect these rapid and distant flights. An individual shorebird may stay only a few days in one spot, so more frequent visits to favored habitats increase encounter probabilities. One day there may be thousands of birds of a half-dozen species. The next day there may be only hundreds, but with the chance to examine fewer birds in greater detail, more species may be identified.

Although there are numerous excellent field guides to North American birds that contain valuable sections on shorebirds, and several books on identification of shorebirds of the world, what has been lacking until now is a book focused on the identification of the shorebirds of North America. The present volume brings together text and photographs to provide the most up-to-date material to facilitate identification of all North American shorebirds.

Acknowledgments

I appreciate greatly the enthusiasm for this project and the patience while I undertook it of my editor at Princeton University Press, Robert Kirk. Nick Lethaby and Steven Mlodinow critically reviewed the entire manuscript and much improved it; you can thank them for their insistence on full treatment of rare species. Steve also kindly loaned me numerous shorebird-identification papers that were not in my own library. My heartfelt thanks to my copyeditor, Elizabeth Pierson, for greatly improving the readability of the book. Others who have given me advice and information about shorebirds since my 1993 book include Don Cunningham, Norio Kawano, Ted Miller, David Roemer, and David Seay. Questions about shorebird occurrence on Barbados were answered by Edward Massiah. Ian Paulsen kindly forwarded many queries over an array of birding listserves. Although not listed in the references, the many authors who wrote shorebird accounts for *The Birds of North America* have my undying gratitude for bringing together so much information in an easily accessible form. I also thank the co-conspirator of my 1993 book, Jim Erckmann, for sharing my enthusiasm for shorebirds over all these years.

Books of this sort are not possible without the prodigious use of museum specimens to check plumage characters and compare species. I have acknowledged the value of numerous collections to my research in *Shorebirds of the Pacific Northwest* and will not duplicate that list here. More recently, I was given additional information or access to specimens by the following curators and collection managers: Keith Arnold (Texas Cooperative Wildlife Collection), Steven Cardiff (Louisiana State Museum of Natural History), Carla Cicero (Museum of Vertebrate Zoology, University of California), Rene Corado (Western Foundation of Vertebrate Zoology), Charles Dardia (Cornell University Museum of Vertebrates), Gene Hess (Delaware Museum of Natural History), Janet Hinshaw (University of Michigan Museum of Zoology), Brad Millen (Royal Ontario Museum), Nathan Rice (Academy of Natural Sciences of Philadelphia), and Paul Sweet (American Museum of Natural History). Wing photos are from the collections of the Slater Museum of Natural History, University of Puget Sound, and the University of Washington Burke Museum; I thank Rob Faucett and Chris Wood for allowing me to take photos of their specimens.

I of course thank the many photographers who made their images available to me and helped shape a book that could be a visual delight as well as an educational tool; without them there would be no book. I used certain photos through the courtesy of Doug Wechsler and Matt Sharp (VIREO), Linnea Hall (Western Foundation of Vertebrate Zoology), and Barb Putnam (BioDiversity Institute). I can only say that acquiring the images for this book has been an interesting combination of logistical nightmare and delight, the latter because of the high quality of the photos and the great generosity and patience of the photographers. Those of you who sent large collections of slides and let me hold onto them until the end deserve my special thanks. As well, I thank all the photographers who submitted photos that were not used.

Above all, my most heartfelt thanks to my wife, Netta Smith, for her tolerance and support while I spent time reading and writing about shorebirds when we could have been doing many other things.

Shorebirds of North America

Introduction

This is a book about shorebird identification, but it should serve also to show just how interesting shorebirds are, as a trip through the photo gallery will confirm. They vary in size and shape and color to a surprising degree, considering they all do more or less the same thing. As the name of the group implies, they are birds that inhabit the shore, where they run or walk or wade or swim and hunt for the small animals, mostly invertebrates, that nourish them. Other birds may be seen at the shore—herons, crows, pipits, gulls, rails—but none of them is a shorebird. Shorebirds are in fact a taxonomic group, consisting of the species in the suborder Charadrii of the order Charadriiformes, the order including also the jaegers and skuas, gulls, terns, skimmers, and alcids (auks, puffins, and their relatives).

Almost all guides to the plants and animals of "North America" include only Canada and the United States, but this one adheres to its name by including the entire continent south through Panama, as well as the West Indies. Although Mexico and Central America are bona fide components of North America, I also refer to them in the text as "Middle America."

Seventy-one species of shorebirds occur regularly (on an annual basis) in North America (Canada, the United States, Mexico, Central America, and the West Indies), although some of them are quite rare on this continent. An additional 12 species occur less regularly and are considered casual. Ten species are known from five or fewer records to date (status from 2002 American Birding Association checklist and subsequent publications) and are considered accidental (Table 1); this can be used as a list of shorebirds you *won't* see in North America. One additional species, Eskimo Curlew, is probably extinct.

Table 1. Shorebirds that are accidental (five or fewer records) in North America.

Species	*No. Records*	*Region*
Greater Sand-Plover	1	California
Little Ringed Plover	3	Alaska
Eurasian Oystercatcher	2	Newfoundland
Black-winged Stilt	2	Alaska
Marsh Sandpiper	2	Alaska
Slender-billed Curlew	1	Ontario
Jack Snipe	5	Alaska, Washington, California, Labrador, Barbados
Pin-tailed Snipe	2	Alaska
Collared Pratincole	1	Barbados
Oriental Pratincole	2	Alaska

All species are given full treatment in the text, as I have noticed that birders appreciate more, rather than less, information on species covered in identification books, even or perhaps especially the rarest ones. These 94 species represent 42 percent of the world's shorebirds (Table 2).

Table 2. Shorebird diversity: number of species in the world and in North America, and within North America, world percentage and number of species that are regular, casual, accidental, and extinct.

				In N.A.				
Family	*Group*	*World*	*N.A.*	*World %*	*Reg.*	*Cas.*	*Acc.*	*Ext.*
Burhinidae	thick-knees	9	1	11	1			
Dromadidae	Crab Plover	1	0					
Charadriidae	plovers	66	18	27	13	3	2	
Pluvianellidae	Magellanic Plover	1	0					
Haematopodidae	oystercatchers	11	3	27	2		1	
Ibidorhynchidae	Ibisbill	1	0					
Recurvirostridae	stilts & avocets	10	3	30	2		1	
Jacanidae	jacanas	8	2	25	2			
Rostratulidae	painted-snipes	2	0					
Scolopacidae	sandpipers	89	65	73	51	19	4	1
Glareolidae	pratincoles & coursers	17	2	12			2	
Pedionomidae	Plains-wanderer	1	0					
Thinocoridae	seedsnipes	4	0					
Chionididae	sheathbills	2	0					
Total for all species		222	94	42	71	12	10	1

(Family and world totals synthesized from Monroe and Sibley 1993 and Clements 2000.)

Anatomy

It is very important to know the details of feather anatomy when trying to identify birds and understand plumage descriptions. Patterns of birds usually correspond to anatomical regions, and this seems especially true for shorebirds.

Shorebird bills are about as variable as the bills of any group of birds, probably in large part because so many of them probe into the substrate, and substrates vary from shallow to deep and hard to soft, so there are many opportunities to specialize in feeding methods. The simplest variable is length, with the shortest-billed birds picking prey from the surface and the longer-billed ones probing into the substrate for it. Bills with droopy tips must give their owners some advantage in picking invertebrates from deep in the mud, the curved bills of curlews may be specially adapted for plucking crabs and worms from burrows, the upturned bills of avocets work best for swinging through the plankton soup in alkaline waters, oystercatchers' chisel-like bills pry mollusks from rocks or snip open shells, and the spoon of the Spoon-billed Sandpiper must be even more specialized, although poorly understood. In any case, these special bills provide terrific field marks for the birds that bear them.

Shorebird legs vary similarly in length, if not in shape. Short-legged species run or walk on sand or mud or rocks, whereas long-legged species typically wade in shallow water. Water too deep for wading has its specialized shorebirds, the short-legged swimming phalaropes. It is important to realize that the prominent joint in a bird's leg is its ankle joint—the joint between the tibia (lower leg) and tarsus (part of the foot)—not its knee joint; a bird's knees do not bend backward!

Shorebird wings are for flying, which would seem to constrain their size and shape greatly. Indeed, most shorebird wings are relatively narrow and pointed, typical of high-speed flyers. Long-distance migrants often have wings of this shape, and shorebirds are among the champions in the extent of their migrations. In addition, they also have daily movements, traveling between feeding grounds and roosts with the tidal cycle, that necessitate rapid and efficient flight. If that weren't enough, they feed in the open and are much at risk from falcons, and their rapid flight, especially to form flocks quickly and efficiently, is probably an asset in escaping from such predators. For some reason not immediately apparent, stilts have the most pointed wings, whereas oystercatchers and lapwings have somewhat rounded wings. At the other extreme are the very rounded wings of woodcocks. In woodcocks, as in grouse, short wings with stiff flight feathers facilitate a rapid takeoff from a potential ground predator in dense woodland.

Anatomical Terms

See Figs. I.1–I.4 for basic shorebird anatomy.

Plumage Variation

One reason museums have such large collections is that organisms vary tremendously, and a large range of specimens is necessary to illustrate that variation. Books can never show it all, but observers who watch for it carefully should be able to detect much of it, at least in common species. It is exciting to realize that this variation, brought about by genetic differences, is the stuff on which natural selection works and is thus significant in the evolution of all shorebirds.

There are five major and at least three minor sources of variation in bird plumages, described below, and shorebirds exhibit them all. Of course, plumage is not all that varies; size and proportions also vary, as do behavior and many attributes of a bird that we cannot see. This variation does not occur to confuse the birder, but no one will deny it has that effect!

Seasonal Variation

A great majority of North American shorebird species show seasonal variation in plumage, with distinct breeding and nonbreeding feather coats. Breeding plumage is often a convenient mixture providing camouflage from above (important in the open breeding habitats of many species) and display colors from below (many species have flight displays). Nonbreeding plumage provides camouflage on the typical wintering habitat, and the substrate on which a species usually forages can be predicted from the coloration of its upperparts. Ventral color does not function much in camouflage, so most nonbreeding plumages feature largely white underparts, as unpigmented

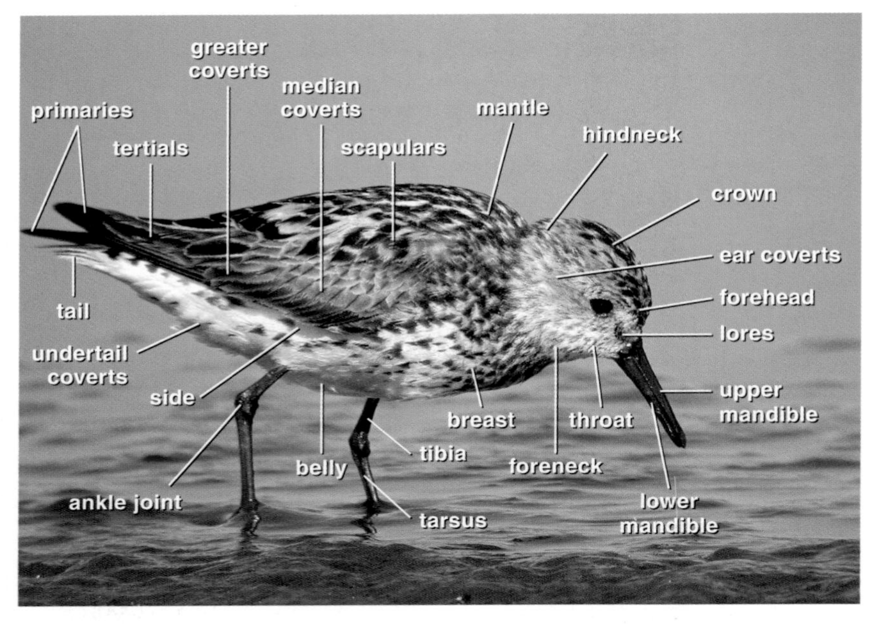

1.1 Anatomy. Western Sandpiper. Padilla Bay, Washington, USA, Jul 2002 (Stuart MacKay).

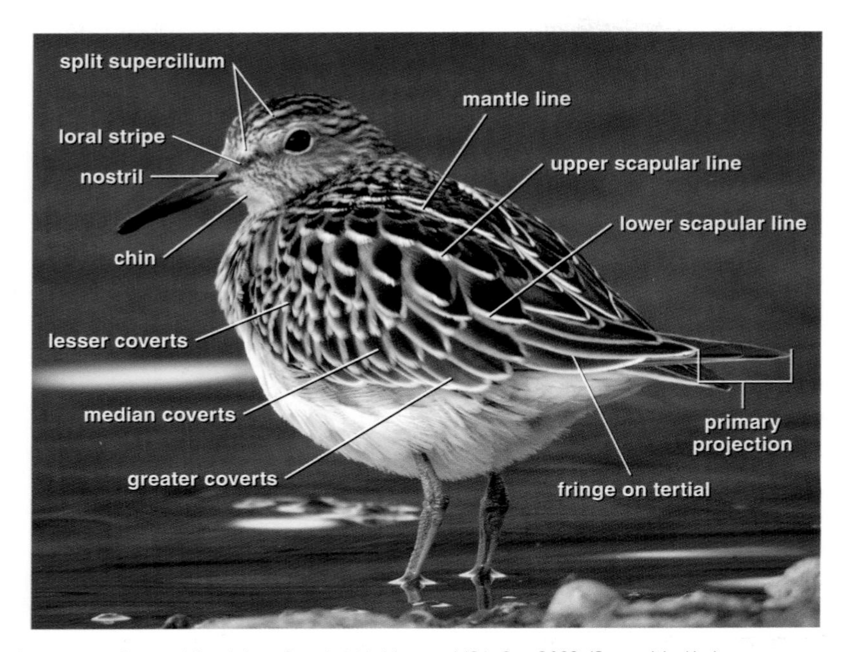

1.2 Anatomy. Pectoral Sandpiper. Seattle, Washington, USA, Sep 2002 (Stuart MacKay).

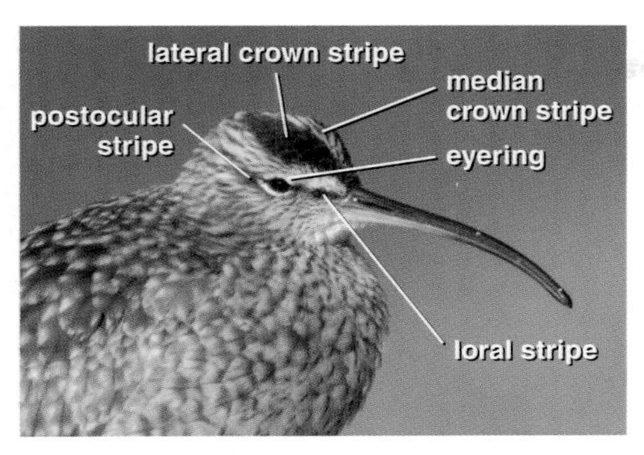

I.3 Head anatomy. Whimbrel. Tokeland, Washington, USA, Nov 2002 (Stuart MacKay).

I.4 Flight anatomy. Ruddy Turnstone. Fort Myers, Florida, USA, May 2003 (Wayne Richardson).

feathers are easiest to produce. Breeding birds may show a mixture of bright (breeding) and dull (nonbreeding) feathers, especially among tertials and coverts. The species that do not change plumage are either large and showy (oystercatchers, stilts) or breed and winter in habitats that are very similar (Buff-breasted and Upland sandpipers). Less well publicized is that bare-part colors vary seasonally in quite a few shorebirds, for example Wilson's Phalarope, with gray legs in summer and yellow legs in winter, and Red Phalarope, with a mostly yellow bill in summer and mostly black bill in winter.

Sexual Variation

In most shorebird species, males and females look identical or extremely similar and are considered monomorphic. In some species, however, the breeding plumages of the two sexes are sufficiently different that they can be identified in the field, and these species are considered sexually dimorphic. The most obvious differences occur in the Ruff, phalaropes, Black-necked Stilt, Ruddy Turnstone, three of four godwits, and *Pluvialis* plovers. Species with less obvious plumage differences that nevertheless usually allow sex distinction include Northern Lapwing, many *Charadrius* plovers, Spotted Redshank, and Curlew Sandpiper. In some shorebirds, there is sufficient sexual variation in bill size and/or shape (presumably representing some difference in feeding ecology) that the bill can be used as a sexual identifier; this also is a mark of sexual dimorphism. Curlews, godwits, avocets, and Western Sandpiper fall in this category. In other groups, there is a slight average difference in the size of the bill and/or entire bird, not usable to sex birds in the field but sometimes helpful in birds in the hand that can be measured; the size differences are surely significant to the birds. Sexual variation in body size is great enough in jacanas, Ruff, and Pectoral Sandpiper that the sex can be distinguished in any plumage.

In the species accounts that follow, "slight" dimorphism indicates an average difference between the sexes that could be visible at close range and would allow sex recognition in the field, but there is overlap, with many individuals not sexable. "Moderate" dimorphism indicates a consistent difference between the sexes that should allow all individuals to be sexed in the field. "Great" dimorphism indicates the sexes look very different.

Age Variation

Shorebirds hatch as precocial downy chicks, not covered in this book but well described by Baicich and Harrison (1997). The first "full-sized" plumage is held by juveniles. The technical term used as an adjective for this plumage has long been "juvenal," but for the sake of simplicity I use the word "juvenile" for both plumage and stage. In most shorebirds, juveniles are recognizably different from adults in either breeding or nonbreeding plumage. During fall migration, when adults may be in either of these plumages or molting between them, juveniles stand out as different. Even in species in which there is minimal difference between juvenile and adult plumages, juveniles can be recognized by their unworn feathers at that time of year. Bare-part colors can also differ in juveniles, usually being more like those of adults in nonbreeding than breeding plumage. Size variation occurs in some groups, such as godwits, curlews, and oystercatchers, in which the bill takes a few months to attain full length. By midwinter, adults have molted into nonbreeding plumage and juveniles into first-winter plumage, but some of the feathers of the juveniles are retained and, a few months older than those of the adults, are more worn than those of the adults, as well as showing juvenile patterns. Worn retained tertials often furnish a clue to immaturity into the following spring, as do primaries that are more worn than those of adults.

In some species, especially those that take more than a year to reach sexual maturity, immatures are recognizably different from adults, often in a plumage in between adult breeding and nonbreeding. Hormonal variation may cause plumage variation at this stage, the difference merely because some individuals underwent an extensive molt and others didn't; or conversely, all individuals molted thoroughly in spring, but the proportion of replaced feathers that corresponded to breeding plumage varied greatly. Some immatures migrate north in spring and stand out by being in nonbreeding plumage in flocks of brightly colored breeders. These individuals apparently do not go all the way to the breeding grounds but spend the summer somewhere between their wintering and breeding destinations. Other immatures merely remain on their wintering grounds. These "oversummering" birds occur in most shorebird species and may turn up anywhere, although they obviously are more to be expected at lower latitudes. Hunting for summering shorebirds is an interesting and educational pastime.

Geographic Variation

Different populations of a shorebird species may become isolated geographically, after which they may diverge from one another genetically and may be sufficiently distinctive to be formally recognized as subspecies. This happens in shorebirds that live, for example, on different coasts of a continent or different regions of the Arctic and Subarctic. Often the differences in populations arise as a consequence of their very different wintering areas and migration routes. For example, Red Knots and Short-billed Dowitchers in North America occur in three populations, each of which has been recognized as a named subspecies. Shorebird subspecies may be quite distinct from one another, as in European- and American-breeding Whimbrels, or they may be sufficiently similar that individual birds aren't always identifiable to subspecies in the field. In most cases birds in breeding plumage are more distinctive than those in juvenile or nonbreeding plumage. Size is also a characteristic of populations, and some subspecies differ at least slightly in size, the most extreme variation being in the Dunlin, in which breeding birds from Greenland and western Europe are perceptibly smaller than those from North America, and populations from Scandinavia and northern Russia are intermediate. Pribilof-breeding Rock Sandpipers are bigger than their relatives on all sides, and it would be interesting to know why. Western Willets are slightly larger than eastern ones, and some observers have claimed to be able to distinguish them by size. I know of no case of geographic variation in bare-part colors.

Subspecies are mentioned in this book only if more than one of them is known to occur in North America. Much attention is given to subspecies in the species accounts and captions—perhaps too much—but it is both interesting and important to conservation to know about these distinctive populations. Almost by definition, subspecies intergrade with one another, and distinction among them in the field is often not possible. The publication by Engelmoer and Roselaar (1998) on geographic variation in shorebirds should be read by anyone with an interest in this topic. Some of the subspecies they define are based on small differences in size in different populations and cannot be distinguished in the field, but they are mentioned in the species accounts.

Individual Variation

Individual variation is a significant component in the issue of shorebird identification. Paintings and photographs in field guides rarely show more than one individual of a given plumage. A few variants may be shown of the highly variable male Ruff in breeding plumage, but the actual variation greatly exceeds that depicted. With the exception of male Ruffs, both variable and vivid because of their complex promiscuous mating system, the most startlingly colored species (oystercatchers, stilts, and avocets) are usually the least variable. You've seen one, you've seen them all. But the more subtly colored species, with bars and stripes and spots and squiggles, vary in the size, number, and placement of those complex markings, even to some degree in their base color. Look at a flock of Red Knots in breeding plumage to see a striking example of this variation.

Some of the variation apparent in a shorebird flock in "full breeding plumage" in spring may be caused by the presence of more than one year class, for example immatures that have not molted into full breeding plumage in their first spring. Other sources of variation in such a flock include molting schedules, which vary among individuals depending on where they winter and the flux of their hormones. Almost any flock in April may contain a wide array of different-looking individuals because of this. But the variation between two individuals also plays a part, and a close look at dowitchers, knots, yellowlegs, or stints shows variation in degree and intensity of spotting and barring, width of feather edgings, and other small differences that—spread over an entire bird—can make two individuals look so different that one could conclude they were different species.

Feather Wear and Fading

Feather wear contributes surprisingly much to the variation in shorebird plumages. Feathers that are molted in fall may be quite worn and faded by the next spring, and feathers molted in spring may be similarly worn and faded by late summer. Fading is probably exacerbated by intense tropical sun and, as well, the lower but persistent sun of the arctic summer, especially at high altitudes. Surfbirds nest in Alaska and Yukon mountains, and the bright rufous colors of May usually have faded dramatically by July. Juveniles of numerous pale-breasted species are washed with bright buff below, that color fading quickly during southbound migration, often leaving not a trace. *Calidris* sandpipers, dowitchers, and phalaropes are all good examples of this phenomenon. Not only does fading occur, but physical wear takes the tips off feathers, so that a bird with dark-based and pale-tipped feathers becomes increasingly dark as the pale tips wear off. Oddly, the reverse doesn't happen; no shorebird gets paler because of the wearing off of dark feather tips.

Unusual Plumages

Unusual individuals are sometimes encountered that exhibit variation outside the norm. Partial or complete albinism can occur in any animal, and occasionally one sees shorebirds that have patches of white feathers (partial albinos), that are entirely white with pale bill and legs and red eyes (complete albinos), or that are generally paler all over but not white (leucistic). The white Sanderling and Least Sandpiper shown as examples of unusual plumages in Figs. I.5 and I.6 are best considered leucistic. Complete albinos, lacking normal eye pigments and conspicuous to predators,

I.5 Leucistic Sanderling. This presumed adult (with two other adults) almost completely lacks feather pigments but shows creamy tints often found in leucistic birds. Not a true albino, as eyes, bill, and legs are pigmented. Mie Prefecture, Japan, Oct 1995 (Shinji Koyama).

I.6 Leucistic Least Sandpiper. Like Sanderling, not a true albino. Identifiable by short wings and primary projection, yellow legs. Probably juvenile, from unworn feathers in August. Conneaut, Ohio, USA, Aug 2003 (Robert Royse).

rarely survive long enough to be observed by a birder. I have seen a pure white, but not red-eyed, Dunlin and an unusual Western Sandpiper, the latter normally colored except for almost entirely white wings, a striking bird in a wheeling flock.

Hybrids

Shorebirds are not renowned for their degree of hybridization, but then again hybrids between closely related species might be difficult or impossible to detect, for example between dowitchers or golden-plovers. With the most notoriety is "Cox's Sandpiper," a hybrid between Curlew and Pectoral sandpipers that is of regular occurrence in winter in southern Australia (see Figs. I.7 and I.8). Its field marks are discussed under Pectoral Sandpiper. "Cooper's Sandpiper," known from one specimen and a few probable sightings, is rather similar, its parentage also among the larger *Calidris* sandpipers. Juveniles photographed in Massachusetts and Japan were presumably one or the other of these two hybrid types, although the Japanese bird in particular had some odd characteristics, especially the very short primary projection. North American birders should be aware of these hybrids, and of a few others that have been suggested within *Calidris* and its close relatives (Dunlin × Purple Sandpiper, White-rumped Sandpiper × Pectoral Sandpiper, White-rumped Sandpiper × Buff-breasted Sandpiper, Baird's Sandpiper × Buff-breasted Sandpiper, Little Stint × Temminck's Stint), when an unidentifiable sandpiper appears in the spotting scope (see Fig. I.9).

I.7 Juvenile *Calidris* hybrid, possible Pectoral Sandpiper x Curlew Sandpiper ("Cox's Sandpiper"). Much like Pectoral Sandpiper but with a longer, more slender, all-black bill and more subdued breast pattern, streaks not extended in center. With Semipalmated Sandpipers. Duxbury Beach, Massachusetts, USA, Sep 1987 (Bruce Sorrie/VIREO).

I.8 Juvenile *Calidris* hybrid, possible Pectoral Sandpiper x Curlew Sandpiper ("Cox's Sandpiper"). Compared with Massachusetts bird and found to be very similar, except for apparently shorter wings (not surpassing tail), which may indicate different ancestry. Ibaraki Prefecture, Japan, Aug 2001 (Masao Fukagawa).

I.9 Juvenile *Calidris* hybrid. Much like Buff-breasted Sandpiper but with more heavily streaked breast and slightly different head pattern. May be a hybrid between that and White-rumped Sandpiper; with late flock of White-rumped, and shaped like it, including bill. Avalon Peninsula, Newfoundland, Canada, Nov 2003 (Bruce Mactavish).

I.10 American Oystercatcher x Black Oystercatcher hybrid. Occasional hybrids between these two species where their ranges meet in Baja California produce birds with variably marked underparts. Palos Verdes, California, USA, Feb 1997 (Don DesJardin).

A hybrid with more easily determined parentage, Black-necked Stilt and American Avocet, has been found a few times in California, and hybrids between American and Black oystercatchers (see Fig. I.10) occur with some frequency where these two species coexist in Baja California.

Molt

Some of the variation in shorebird plumages can be attributed to molt. Birds molting between plumages that are very different (breeding and nonbreeding, juvenile and first winter) can look quite different from any of the illustrations one usually finds in field guides. It is impossible to illustrate all of that variation, but birds in molt are shown frequently in this book to illustrate the ways in which birds can be in transitional plumages.

The molt that occurs in body plumage is seasonally predictable, especially in mature adults that molt into distinct breeding and nonbreeding plumages each spring and fall. These body molts each take up to several months, so they represent a significant investment in a bird's time and energy and a significant period during which it looks "funny." Juveniles typically molt most of their body feathers in southbound migration or after arriving on their wintering grounds in their first autumn. The same birds may molt into an adult breeding plumage in their first spring (often molting later than adults) or may retain an "immature" plumage that looks much like the nonbreeding plumage of adults or something in between breeding and nonbreeding. They will then molt again into a more typical nonbreeding plumage in their second autumn. Wing and tail molts occur somewhat independent of body molt, usually in late spring or summer in first-year birds (their flight feathers, weaker than those of adults, may be extremely worn by that time), then later in fall in second-year birds and at the same time for the rest of their life. Variations on that theme involve mid-winter wing and tail molt in some long-distance migrants.

Shorebird molt occurs in a regular fashion. In fall, the upper scapulars often show the first sign of it. The head and flanks also begin to molt quickly, and body molt becomes general all over the body as it progresses. Often the tertials and some coverts are the last to molt, in many cases being retained onto the wintering grounds and not molted until midwinter or not at all until a subsequent molt in spring. Postbreeding flight-feather (wing and tail) molt may be started on the breeding grounds in a few species, but it is usually arrested during migration in those species and not begun until on the wintering grounds in the majority of species. Most exceptions are relatively short-distance migrants, including avocets and stilts, Killdeer, snipes, and woodcocks, but a few northerly breeding species such as Greater Yellowlegs, Least Sandpiper, and Long-billed Dowitcher also complete flight-feather molt when they stop somewhere along the route.

Identification

Most people consider shorebirds to be identification challenges, and indeed many of them are. Quite a few species are similar in overall size and shape, bills of closely related species can be identical, and a tendency toward cryptic plumage makes most of them gray to brown, streaky and spotted, like seagoing sparrows. Yet the majority of the species discussed here are sufficiently distinct that there should be no problem distinguishing them from all other shorebirds. The most distinctive species are the oystercatchers, avocets, and stilts, but some plovers (e.g., Killdeer) and sandpipers (e.g., Marbled Godwit, Black Turnstone) are just as easily recognizable.

On sighting an unfamiliar shorebird, the attributes to look for first are size (easiest if known species are present for comparison), proportions (especially relative length of bill and legs), overall coloration, and leg color. Then focus on the details of the markings on the head, back, breast, and belly. Many of the details of the back are in fact details of discrete feather groups associated with the wings—the scapulars, coverts, and tertials—and they must be correctly allocated to the right group. Three more attributes important to identification remain: behavior, flight pattern, and flight calls. Identification is made much more certain if all of these characteristics can be noted. Remember that many shorebird field marks are *indicative* rather than *definitive*, and identification often rests on seeing as many of the indicative marks as possible.

One aspect of shorebird identification has been emphasized again and again: know your plumages! Typically birds in breeding, nonbreeding, and juvenile plumages look different from one another, and it is commonplace to see individuals of a species in two or even three of these plumages at certain times of year. It is important, if not imperative, to be able to distinguish these plumages in each species to master identification skills.

Shorebird identification has two things going for it. First, shorebirds are birds of the open, so you can watch them for lengthy periods. Only a small minority of species are skulkers. Although at times the birds may be distant, modern spotting scopes compensate greatly for this. Second, shorebirds are social, so a large proportion of individuals will be seen with others of their own or other species, allowing many different comparisons. Imagine how convenient it would be if all birds came in mixed feeding and roosting flocks, so that Henslow's, Le Conte's, and Baird's sparrows or fall Blackpoll, Bay-breasted, and Pine warblers could be directly compared right out in the open!

For much information about plumages and identification and good illustrations of Northern Hemisphere shorebirds, I recommend the following: *The Birds of the Western Palearctic,* vol. 3, edited by Stanley Cramp and K. E. L. Simmons (Oxford University Press, 1983); *Shorebirds: An Identification Guide to the Waders of the World* by Peter Hayman, John Marchant, and Tony Prater (Houghton Mifflin, 1986); *The Facts on File Field Guide to North Atlantic Shorebirds* by Richard J. Chandler (Facts on File, 1989); *Shorebirds of the Pacific Northwest* by Dennis Paulson (University of Washington Press, 1993); *Photographic Guide to the Shorebirds of the World* by David Rosair and David Cottridge (Facts on File, 1995); and the many species accounts in *The Birds of North America* series (published by the Academy of Natural Sciences and the American Ornithologists' Union).

Behavior

A knowledge of shorebird behavior when the birds are away from their breeding grounds is important for finding them, for understanding what they are doing when you find them, and for identifying many species. Shorebirds are active birds, and because most of them are creatures of the open, their behavior can be easily observed.

Much of the behavior of shorebirds (at least those on salt water!) revolves around tidal cycles. Shorebirds feed when the tide is out and roost when it is in, but there is much variation around these two alternatives. Typically, hungrier birds begin feeding most quickly as the tide recedes, and satiated birds return to roost earliest, so there is a considerable spread in when birds move to and fro. The best times to see large numbers of shorebirds from shore are obviously when the tide is just a little way out or almost all the way in, forcing the birds nearer to the observer. In many situations, feeding birds may be far away when mudflats are extensive and humans walking onto them is untenable. After roosting for many hours, shorebirds are usually in a hurry to begin feeding, so as soon as the tide starts out, they will head for the appropriate exposed substrates and be preoccupied with feeding.

Unlike us, shorebirds typically have two meals a day—one long feeding bout at each tidal cycle. When one of the low tides is at night, many species feed at that time, and nighttime low tides are often indicated by the flight calls of shorebirds moving about. Although not on tidal cycles, Killdeers often call as they fly around farmlands and suburbs at night, Wilson's Snipes can be heard winnowing after dark, and American Woodcocks display at dusk, then go on to feed at night.

Another opportune time for shorebird observation is while the birds are at roosts. Often roosts are on inaccessible sites such as islands (fortunately for the birds), but some of them are accessible to birders. The birds are spooky at that time, so every precaution should be taken not to disturb them. If the nearer birds start shuffling away, it's time to stop or retreat, as a few birds taking flight might stimulate the entire flock to do so, leaving the area and spoiling both your viewing and their rest and relaxation. The roosting period allows them to relax some of their vigilance against predators, because the sheer numbers of birds present provide more watchful eyes, more confusion to the predator, and a much lower probability that a particular bird will be a target. Even freshwater shorebirds "roost," often just stopping where they are for a few hours of snoozing, sometimes gathering in flocks to do so.

Shorebirds vary greatly in their feeding behavior. Two basic foraging modes are the plover mode—running, stopping, and running again, with prey capture at the end of a run—and the sandpiper mode—moving steadily over the substrate, whether land or water, rocks or mud, to forage. The plover mode is suited to smooth and open substrates such as mudflats, sandy beaches, and open grassland. Sandpipers, by contrast, live in all shorebird habitats and exhibit much variation in foraging techniques, with some species moving rapidly and searching visually, others moving slowly and hunting tactilely by probing beneath the surface. Rock-foraging sandpipers have short bills and short, thick legs; sand foragers may have slightly longer bills for picking and probing; and mud-foraging sandpipers run the gamut of bill length, short for picking and increasingly long for probing to greater and greater depths. Mudflat species often have small webs between one or two pairs of front toes. Other sandpipers—for example, yellowlegs—forage in water and have relatively long legs for wading and slender bills for picking prey from the surface. Stilts, in another family, are the epitome of this foraging mode. Their near relatives, the avocets, also feed in water but do so by a back-and-forth scything motion, straining out by touch the same invertebrates that the stilts see and grab singly. Finally, phalaropes have relatively short legs with lobed toes that propel them through the water like little ducks, and they often twirl in place to draw their prey closer. Bear in mind that shorebirds don't always play it by the book; Whimbrels have been seen catching butterflies on lawns in Panama.

Most North American shorebirds, no matter their habits, have flight displays—of great value in species that live in open country. Surprisingly, many of those that breed in forested wetlands perch readily in trees and sing from them, but all those species also have flight displays. Songs vary from rapid and buzzy to languid and sweet, but all are loud and far carrying. Stilts, avocets, oystercatchers, jacanas, and some plovers and sandpipers lack flight displays, and Ruffs and Buff-breasted Sandpipers make up for that lack in the fanciness of their ground displays.

Vocalizations

Vocalizations are important clues to shorebird identification and should be learned if at all possible. Most of the ones we hear in migration and winter are flocking calls and flight calls, probably with the similar function of calling together individuals of the species to flock for flights to and from roosts and as an escape tactic when confronted by aerial predators such as falcons. Some birds make quieter calls while feeding, which may be contact calls much as one hears from a flock of kinglets and chickadees as they move through the woods. Vocalizations vary from jarringly loud, probably typical of "sentinel" species that are alert and often warn of the presence of predators, to quiet and virtually lacking (we don't know why some species are so quiet).

During spring, when hormone levels are increasing, many species start to give parts of their songs, and it is exciting to hear these very different sounds coming from a passing flock. The same hormones may cause remnants of song to be heard in fall. To really appreciate shorebird vocalizations, the birds must be visited on their breeding grounds, especially at high latitudes, where so many of them breed. In the breeding season, shorebirds adopt a strategy opposite that of winter—most of them become hyperconspicuous visually and vocally, no different from a Baltimore Oriole or Redheaded Woodpecker. Then the complexity of their songs can be appreciated, as well as an entire array of additional vocalizations between adults and chicks. Most of the sounds made by shorebirds are of course vocal, but a few species produce mechanical sounds in flight with their wings and tails.

Distribution

Range maps are not included in this book, in part so all the space can be used for photos and text but also because all current field guides have largely accurate range maps that can be used in conjunction with this book. Although a few of the species treated here are tropical residents without much tendency to roam, shorebirds as a group are among the most motile of organisms, commonly occurring outside their "usual" distribution. Wintering birds may be common at one latitude and steadily less common with increased latitude, and drawing a line that implies "does not winter north of here" is difficult. It is even more difficult to draw range maps that adequately show the complex migratory pathways taken by many shorebird species. In some species, adults take entirely different routes in spring and fall, and in some species juveniles may take yet another route, so an individual shorebird may use three different migratory pathways during its lifetime. These are briefly described in the species accounts below when they are not merely broad movements all across the region between breeding and wintering grounds. Some species migrate by short hops, others by longer movements between feeding areas, the most extreme using only one or two "staging areas" between the two ends of their journey. These strategies have been described as "a hop, a skip, and a jump!"

Because of their great latitudinal migrations, strong flight, and potential for displacement by east-west winds, it should follow that a wandering shorebird could turn up just about anywhere. This is why most North American shorebirds have been recorded in just about all Canadian provinces and American states. It is why most of the high-latitude Eurasian shorebird species are included in this book, because many of them wander to North America on an annual basis, far from their Australian, Asian, or African wintering grounds. It is also why Siberian species such as Red-necked Stint and Sharp-tailed Sandpiper have been recorded all the way to the American Atlantic coast more than once. Apparently they oriented at right angles to or even exactly opposite their usual migratory pathway as they set out on fall migration. And that is why birders keep a sharp eye out for unexpected species. However, it still must be kept in mind how unlikely it is for an individual birder to come across a very rare shorebird, and sightings of what are considered rare species should be carefully documented. Rare shorebirds are only *possible*, not probable. In the species accounts in this book, the term "vagrant" is used for a bird that occurs well outside its usual range.

Conservation

Shorebirds, like many other migratory birds, are at risk because so many of them depend on multiple habitats and homes. Protecting a typical North American migratory shorebird involves protecting its breeding grounds, its wintering grounds, and the several sites at which it stops over in migration to feed and replenish fat supplies for further migration. Each of these areas is critical to the bird, and such a chain can develop a weak link if even one of them is threatened. Human activities over a huge latitudinal extent can affect the fate of a single shorebird species, so oil drilling in the Arctic National Wildlife Refuge, construction in the Bay of Fundy to harness tidal power, dredging and filling in San Francisco Bay, recreational development in Florida and Mexico, illegal hunting in Colombia, and oil drilling in Argentina may all conspire against a typical shorebird. And that is just for a coastal migrant. Interior wetlands that are important to many different shorebirds are being lost at a more rapid rate.

The areas of greatest shorebird concentrations are typically the areas of greatest production of invertebrate life, so as such they are important to many organisms other than shorebirds. Shorebird concentrations can thus be apparent indicators of significant marine and freshwater wetlands, without the necessity of sampling their prey. Such sampling is vital to understanding shorebird biology, but the shorebirds themselves will contribute to their own conservation by telling the story of habitats to be guarded against exploitation and development.

Fortunately, much effort is being expended to protect important shorebird areas, and the Western Hemisphere Shorebird Reserve Network (WHSRN) has been established to accomplish this in the New World. The United States Shorebird Conservation Plan (Brown et al. 2001) presents information about shorebird populations and efforts to conserve them. Rough estimates of breeding and migrant shorebird populations in the United States and Canada are given in Table 3. The totals may be surprising to some birders who have never seen one or more of the species of which there are thought to be over a million individuals on our continent! But the species at the lower end of the list should be our greatest concern, as any bird with a small population is vulnerable to both environmental and human perturbations of its habitats.

Conservation depends on knowledge, and birders can furnish that knowledge by participating in shorebird surveys. Many of them are of local origin, but the International Shorebird Survey (ISS) was begun by the Manomet Center for Conservation Sciences for the purpose of pulling together such surveys from all over the world. Further information on both the WHSRN and ISS can be found on the World Wide Web.

Table 3. Estimated populations of breeding and migrant shorebirds in Canada and the United States. (Degrees of confidence vary greatly; from Morrison et al. 2001.)

Species	*Population*	*Species*	*Population*
American Woodcock	5,000,000	Semipalmated Plover	150,000
Semipalmated Sandpiper	3,500,000	Black-necked Stilt	150,000
Western Sandpiper	3,500,000	Spotted Sandpiper	150,000
Red-necked Phalarope	2,500,000	Rock Sandpiper	150,000
Wilson's Snipe	2,000,000	Greater Yellowlegs	100,000
Wilson's Phalarope	1,500,000	Bar-tailed Godwit	100,000
Killdeer	1,000,000	Black Turnstone	80,000
Red Phalarope	1,000,000	Surfbird	70,000
Dunlin	850,000	Whimbrel	57,000
Least Sandpiper	600,000	Hudsonian Godwit	50,000
Lesser Yellowlegs	500,000	Solitary Sandpiper	25,000
Long-billed Dowitcher	500,000	Long-billed Curlew	20,000
American Avocet	450,000	Pacific Golden-Plover	16,000

Table 3. *(cont.)*

Species	*Population*	*Species*	*Population*
Red Knot	400,000	Snowy Plover	15,700
White-rumped Sandpiper	400,000	Purple Sandpiper	15,000
Pectoral Sandpiper	400,000	Buff-breasted Sandpiper	15,000
Upland Sandpiper	350,000	Common Ringed Plover	10,000
Short-billed Dowitcher	320,000	Wandering Tattler	10,000
Sanderling	300,000	Bristle-thighed Curlew	10,000
Baird's Sandpiper	300,000	Mountain Plover	9,000
Willet	250,000	Black Oystercatcher	8,900
Ruddy Turnstone	235,000	Wilson's Plover	6,000
Black-bellied Plover	200,000	Piping Plover	5,800
Stilt Sandpiper	200,000	American Oystercatcher	3,600
Marbled Godwit	171,500	Sharp-tailed Sandpiper	3,000
American Golden-Plover	150,000	Eskimo Curlew	<50

Species Accounts

The species accounts follow the order of the American Ornithologists' Union checklist (1998), and they are all presented in the same way, with boldface sections. Each account begins with a brief synopsis of the species, followed by measurements (2.5 centimeters = 1 inch) and brief descriptions of contrasting plumages. When appropriate, subspecies are then listed and discussed. Identification notes include comparisons with similar species, typically only under one species to conserve space. Flight patterns, voice, and behavior are then described. Breeding-ground vocalizations are given only for species that breed in North America. Brief statements on habitat and range conclude each account. The descriptive parts of the species accounts are given in telegraphic style to save space; I hope the reader is not offended.

Measurements given are of course averages, as all species vary. In cases where different populations or sexes differ slightly in average size, I use a midpoint between them. In cases where the difference might be evident in the field, I include measurements for both groups. Weights in particular vary tremendously with annual cycles and the deposition of large quantities of fat, and I tried to get as close as I could to what might be considered an average lean weight. Lengths were taken from specimens I considered typical of a species, as if a bird were laid out with its neck moderately extended; I consider the measurements in many guides, perhaps taken from greatly extended birds, misleading.

Complete plumage descriptions are not given; they can be found in most of the shorebird references listed earlier. For plumages that differ from one another, only the differences are noted. Immatures are not discussed separately from juveniles and adults unless they comprise a distinct plumage type. They can usually be recognized through the winter and spring by their increasingly worn primaries, and often tertials, in comparison with adult birds. Note that all the shorebirds included here have brown eyes, except thick-knees, oystercatchers, and adult stilts.

The brief descriptions of behavior refer to migration and winter, the times when shorebirds are usually seen by most of us. Their breeding behavior is surpassingly interesting but is not described herein, although species such as Killdeer, Spotted Sandpiper, and Wilson's Snipe breed sufficiently widely that most birders can become familiar with basic shorebird breeding behavior by watching them. Habitat descriptions are generalizations, of course, intended to be helpful but by no means complete. Upland species occasionally visit beaches, and coastal species are rarely found inland. Vagrant occurrences of North American shorebirds are commonplace in the Old World but are not included in range statements.

Photos and Captions

Well over 5,000 photos were scrutinized to choose the 534 used here. The photos were chosen to illustrate as many variations in the species as possible; a relatively small number could not be located in the collections of the many photographers contacted. Throughout the period of choosing photos, I attempted to get the best possible illustrations of identification features. I also attempted constantly, with a high degree of success, to find photos that have not been previously published in identification guides. The wandering tendencies of shorebirds could not be better illustrated than by the number of photos used here from outside North America.

The captions are intended to make as many of the photos as possible instructional themselves. Localities and months are included, as I consider these important when looking at photos. Each photo is, like a museum specimen, only a representative of its species, a data point in space and time. In many cases the month dates were taken from the processing date on a slide, meaning the photo could have been taken in an earlier month. I will add a plea here to all nature photographers to label your photos with locality and date, which are just as important as species identification!

Thick-knees (Burhinidae)

This family of nine species, only one of which occurs in North America, is mostly tropical and mostly Old World. They are large shorebirds with quite large heads and large eyes for crepuscular and nocturnal activity, and long legs for covering open ground rapidly. Some live near water, others in upland habitats. The large bill allows them to take quite large insects, and two large beach-dwelling species eat crabs. The ankle joint is perhaps slightly thicker than that of other shorebirds (you can't see birds' knees from the outside).

1 Double-striped Thick-knee

(Burhinus bistriatus)

Only a single individual of this striking and exotic species has wandered north into U.S. territory, but it is locally common in open country of Middle America. Its calls coming out of the darkness are often the first clues to its presence.

Size Weight about 780 g, length 42 cm (16.5 in), bill 5.0 cm, tarsus 11.4 cm.

Plumages Eye bright yellow. Bill black at tip, yellow at base, culmen dark gray. Legs yellow. *Adult* large and brown, streaked with black and with white supercilium and black stripe above it. *Juvenile* similar but somewhat duller and grayer; black crown stripe and supercilium start slightly farther back and latter bordered below at rear with dark mark.

Identification Unique, large shorebird with long legs, big head, heavy bill, and bright yellow eye, very different from any other North American species. Colored for its upland habitat, evenly streaked with darker and lighter brown above, with brown, finely streaked breast and white belly. Broad white supercilium and black stripe above it are distinctive. Stone-curlew (*B. oedicnemus*) of Europe and western Asia seems very unlikely vagrant to North America; considerably smaller and shorter-legged, with different head pattern and white stripe on folded wing.

In Flight Big and brown, with long legs extending beyond tail, narrow wings with fluttery wing beats and conspicuous white patches on both outer and inner primaries.

Voice Calls repeated loud, ringing or cackling notes, rising and then fading away, usually heard at night; may sound froglike. Southern Lapwing calls somewhat similar, heard by day.

Behavior Forages for large insects, walking sedately or running and tipping like plover. Normally nocturnal, especially for feeding, but may be seen during day, even in small groups, when right out in open. Often folds legs, rests on tarsi.

Habitat Grassland, with or without scattered shrubs, and open savanna.

Range Resident from southern Veracruz and Oaxaca south to northwestern Brazil, also on Hispaniola. Vagrant to USA, one winter record from southern Texas.

1.1 Adult Double-striped Thick-knee. Unmistakable large, long-legged, big-eyed shorebird with conspicuous black and white head pattern. Bold white pattern in wings in flight. Hato Piñero, Venezuela, Jan 1998 (Joe Fuhrman).

Plovers (Charadriidae)

Sixty-six species occur worldwide, 13 of them regular in North America plus 5 vagrants. Plovers can be universally recognized by their foraging method, an alternation of running and stopping, with prey detected visually during either the run or the stop. They often run instantly upon prey capture. If no prey is seen after a few seconds, the plover runs again. As befits picking shorebirds, they are short billed, the bill thicker at the base and tapering to a fine tip. The bill is always hard at the tip, unlike in many sandpipers. Most plovers are three toed, but Black-bellied Plover and lapwings have a hind toe.

Lapwings *(Vanellus)*

Lapwings are large plovers with rounded wings and wing spurs; most are tropical residents, but a few extend to higher latitudes and are at least partially migratory. The 23 species worldwide vary greatly, but all are strikingly colored, usually plain earth tones above but often mixed with black and with contrasty head and breast patterns. Some of them have crests, some have conspicuous wattles around the face, and some are iridescent green above. They are all spectacularly conspicuous in flight, many of them with big white wing patches, and most are noisy, truly "sentinel" shorebirds.

2 Northern Lapwing

(Vanellus vanellus)

On the rare occasion when one of these large plovers appears on our shores, it is unmistakable on the ground or in flight. Lapwings are bold and noisy where they normally occur, but they are probably more subdued after an ocean crossing. Surprisingly many have made the crossing, and they are anything but inconspicuous, so observers on the North Atlantic coast should be on the alert for this species.

Size Weight 206 g, length 29 cm (11.5 in), bill 2.4 cm, tarsus 4.7 cm.

Plumages Bill black, legs dull red. Slight plumage dimorphism. *Breeding adult* black on forepart of face and neck, male entirely so, female with white mottling in same areas, also somewhat shorter crest than male. *Nonbreeding adult* like breeding but with white chin and foreneck, even less black than breeding female. Narrow buffy fringes on scapulars and white fringes on black breast feathers produce scaly effect on back and breast. Uniquely among northern shorebirds, also sexual dimorphism in nonbreeding plumage: female duller than male, with shorter crest, browner head, and duller, more olive back. *Juvenile* much like nonbreeding adult but with gray-brown crown, even less black on face, distinctly shorter crest, more heavily scaled upperparts, and gray legs.

Identification No other North American shorebird is similar to this species. Looks black and white at a distance, but at close range iridescent oily green back and purplish tinge to scapulars visible, also rufous undertail coverts, striking black and white face pattern, and long, wispy crest. Far to south, Southern Lapwing bronzy rather than oily green above, lacks rufous undertail coverts of this species, has pinkish red bill and eyes, and shows prominent white wing patches in flight. Confusion seems unlikely, even at a distance, with black and white Black-bellied Plover or largely black male Ruff in breeding plumage.

In Flight Dark above and white below, tail with white base and black tip. As distinctive in air as on ground, with broad wings and almost butterflylike flight, rather slow and with irregular wing beats. White wingtips only wing pattern from above. Primaries longer in males than in females, making outer wing look very broad, and—uniquely—sexual dimorphism in wing pattern, female with broader white tips to primaries. White underparts include underwing coverts.

Voice Flight call a loud, shrill, double note, second syllable higher, giving rise to British name "Peewit" and German name "Kiebitz." Quite vocal, at Killdeer level of overtness.

Behavior Runs and stops in typical plover fashion, often hunting for earthworms in plowed fields. Spreads out to feed, collects in large flocks where common.

Habitat Moist and dry open grassland, meadows, moors, and farmland, often associated with freshwater. May move to mudflats in winter.

Range Breeds all across northern and central Eurasia. Winters south to far northern Africa, Middle East, and southern Asia. Vagrant to Atlantic coast of Canada and northeastern USA, most records from two winter incursions; more rarely in late fall and winter west to Ohio and south to Florida. Also records scattered throughout West Indies from Bahamas to Barbados.

2.1 Breeding adult Northern Lapwing. Unmistakable crested and iridescent green, black, and white plover. Probably male because of mostly black face and long crest. Outer Hebrides, Scotland, Jun 1993 (Mike Danzenbaker).

2.2 Nonbreeding adult Northern Lapwing. Distinguished from breeding by less black on face, buff-tipped scapulars and coverts. Modena, Italy, Dec 1995 (Fabio Ballanti).

2.3 Juvenile Northern Lapwing. Distinguished from adult by more pale color on head and more extensive pale fringes on upperparts. Kohokucho, Shiga Prefecture, Japan, Nov 2002 (Tadao Shimba)

2.4 Breeding adult Northern Lapwing. Black and white plover with broad wings, white-tipped outer primaries. Long inner primaries indicate male. Derbyshire, England, Jun 1988 (Mike Danzenbaker).

3 Southern Lapwing

(Vanellus chilensis)

This species, almost ubiquitous in open country in South America, is well known to country people because of its persistent noisiness and mobbing behavior. It is a prominent feature of savannas in the north and pampas in the south.

Size Weight 285 g, length 30 cm (12 in), bill 3.3 cm, tarsus 7.9 cm.

Plumages Bill tip black; bill base, eyering, eye, and pointed wing spur red; legs dull red. *Adult* with head gray; wispy crest, face, stripe down neck, and breast black; white line separating black and gray of head. Back olive; iridescent bronze on scapulars blends with iridescent green/purple on lesser wing coverts. *Juvenile* like adult but with upperparts fringed buff, face pattern less distinct.

Identification Nothing else like it in its range except Andean Lapwing (*V. resplendens*), with no crest, entire head and breast pale, occurs mostly well up in mountains.

In Flight Unmistakable, with white coverts forming curved band across each wing; tail black, rump and tail base white. High-altitude Andean Lapwing very similar, shorter legs not extending beyond tail (toe tips beyond in Southern). Much smaller Pied Plover (*Hoploxypterus cayanus*) of lowlands has black and white pattern on crown and back.

Voice Series of one- and two-note calls when disturbed, between harsh and musical but always loud and persistent. Like other lapwings, named after its calls ("terotero" in Argentina).

Behavior Runs and stops, spread out to forage. Often feeds at night. Single or in pairs or small groups, not very social. Noisy and persistent mobber, as much as any shorebird.

Habitat Moist to wet grassland, pastures and agricultural lands, and edges of marshes and lakes.

Range Resident throughout almost entire lowlands of South America and up to middle elevations in mountains. Casual north to Costa Rica and Barbados and appears to be slowly expanding northward.

3.1 Adult Southern Lapwing. Large plover with vivid pattern, pink to red eyes, bill, and legs. Long black crest characteristic. Calabozo, Venezuela, Sep 1978 (Dennis Paulson).

3.2 Adult Southern Lapwing. Broad wings typical of lapwings, with large white wing patches and black and white tail from above. Conspicuously black and white from below. Calafate, Argentina, Nov 1995 (Dennis Paulson).

Tundra Plovers *(Pluvialis)*

The four species of this genus, all of which occur in North America, are large plovers, exceeded in size in their family only by some of the lapwings. All have largely spangled upperparts and mostly black underparts in breeding plumage, but in nonbreeding and juvenile plumages they are subdued brown, gray, and golden above and white to gray below with streaks and bars. This is far more plumage change than in any other group of plovers, and the black underparts may be an adaptation making them more conspicuous in their far-flung flight displays on the breeding grounds. Black-bellied Plover, a beach inhabitant, is the grayest species and has the flashiest flight pattern, whereas the three species of golden-plovers are upland dwellers and are brown with plain wings and tail. Interestingly, the details of juvenile plumage are reminiscent of the same plumage in tringine sandpipers.

4 Black-bellied Plover

(Pluvialis squatarola)

The largest of our native plovers, Black-bellied (or Grey Plover outside the Americas) is also one of the wariest, flushing at great distances with noisy whistles when humans appear on the scene. Its flight call is one of the more evocative of shorebird sounds, and it remains sufficiently common that most North American birders will have the opportunity to hear it at some time during the year.

Size Weight 210 g, length 28 cm (11 in), bill 3.0 cm, tarsus 4.4 cm.

Plumages Moderate plumage dimorphism. Bill and legs black. *Breeding adult* looks black and white from a distance, with much black below. Male with pale crown, vividly barred back, and solid black underparts except undertail. Female varies from almost as brightly marked as male to much duller, with mostly brownish upperparts and much white in black of underparts. Outer undertail coverts often more heavily marked than in males. Even tail dimorphic, white with narrow dark brown to black bars in male and often brownish, with brown bars in female. *Nonbreeding adult* with legs medium to dark gray. Rather plain gray-brown above and white below, most conspicuous markings darker brown feather bases; tertials with alternating darker and lighter dots or notches along edges. Breast and sides vary from pure white through grayish to fairly heavily mottled with brown. Foreneck with brown streaks or, rarely, heavily flecked with dark brown on neck, breast, and sides. *Juvenile* also with legs gray, superficially like nonbreeding adult but distinguished by more distinctly pale-spotted upperparts and heavily streaked breast. Spots on upperparts may be rather yellow at first, quickly fading to white. After autumn molt, still recognizable through winter by retained dark tertials with ragged edges corresponding to white dots. Immatures in spring are often in nonbreeding or mixed plumages.

Subspecies Birds breeding in Eurasia and Alaska are considered *P. s. squatarola*; Alaska birds average larger than Canadian-breeding *P. s. cynosurae*. Determination of wintering grounds and migration routes of these populations needed.

Identification Most likely to be mistaken for a golden-plover but differs in being paler overall in all plumages, with no yellow tones except on fresh juveniles, which can have pale spots on upperparts tinted with that color. Also larger than any golden-plover, with distinctly larger bill. Streaks on underparts of juveniles stop at legs, no bars at all, whereas golden-plovers marked behind legs, usually with bars. New World golden-plovers have gray axillars and underwings and at least some black on undertail coverts in breeding plumage, whereas Black-bellied has black axillars, white underwings, and white undertail coverts.

In Flight Stripe winged and white tailed, this plover stands out by its black axillars at all times of year (because of this, black of underparts in breeding plumage projects farther out under wings than in golden-plovers). Birds in breeding plumage black beneath, from throat to lower belly, but with white underwings and undertail coverts. In flight Black-bellied more likely to be mistaken for Red Knot (or very rare Great Knot) from above but larger, with shorter bill and more conspicuous wing stripes; black axillars eliminate all other species.

Voice Flight call a beautiful, plaintive, whistled *whee-er-eee* or *klee-er-ee* given frequently by flying or alert birds. No other bird sounds anything like this species, with exception of Bristle-thighed Curlew, call of which is abrupt rather than slurred. Breeding-ground calls include sharp, high-pitched, three-noted whistles during display flight.

Behavior Runs and stops like a typical plover. Spreads out across feeding habitat; often aggressive to others of same species but may roost in large dense flocks. Usually sleeps with bill exposed. Sometimes feeds at high tide by walking in shallow water, picking prey from surface—quite unploverlike.

Habitat Dry tundra for breeding. Coastal beaches, estuaries, and flooded fields during migration and winter; migrants also on lake shores.

Range Breeds scattered across Alaskan and Canadian Arctic, also across Eurasia. Winters from southern British Columbia and New England south through Caribbean on coasts to Peru and northern Brazil, rarely to southern South America. Also in Europe, Africa, Asia, and Australia. Migrants throughout, more common on coasts but widely in interior, where most common on northern Great Plains and around Great Lakes.

4.1 Breeding male Black-bellied Plover. Vivid black and white above, entirely black below except undertail coverts. White crown and virtual lack of brown on upperparts indicate male. Nome, Alaska, USA, Jun 1998 (Brian Small).

4.2 Breeding male Black-bellied Plover. Black and white barred tail also characteristic of sex. White underwing and black axillars distinguish from golden-plovers. Nome, Alaska, USA, Jun 1998 (Brian Small).

4.3 Breeding female Black-bellied Plover. Usually much duller than typical male, with much more uniform upperparts and much white mixed with black below. San Diego, California, USA, May 1979 (Anthony Mercieca).

4.4 Adult Black-bellied Plover. Photographed in late November, this individual has retained much of its black underparts while entirely in nonbreeding plumage above. Dark markings on underparts are often last sign of breeding plumage in fall. Black and white tail indicates probable male. Santa Barbara, California, USA, Nov 1995 (Larry Sansone).

4.5 Nonbreeding Black-bellied Plover. Very plain above, streaks minor part of pattern on underparts. Tiny hind toe characteristic of species (not present in most plovers). Ventura, California, USA, Jan 1996 (Brian Small).

4.6 Juvenile Black-bellied Plover. Pale notches (can be yellow early in season) all along dark feathers of upperparts and heavily streaked underparts diagnostic of this plumage. Distinguished from juvenile golden-plovers by bigger head and bill, predominance of streaks below. Conneaut, Ohio, USA, Oct 2001 (Robert Royse).

4.7 Immature Black-bellied Plover. By this date, much like adult but retained worn juvenile coverts and tertials indicate first year. Ventura, California, USA, Dec 1995 (Brian Small).

4.8 Breeding male Black-bellied Plover. Flight pattern includes conspicuous white wing stripe and white tail in any plumage. Unmistakable black and white pattern in this plumage, much plainer in nonbreeding plumages. Fort Myers, Florida, USA, May 2001 (Wayne Richardson).

4.9 Nonbreeding Black-bellied Plover. White underparts with contrasty black axillars diagnostic. Two Red Knots in upper part of flock. Orange County, California, USA, Dec 1999 (Brian Small).

5 European Golden-Plover

(Pluvialis apricaria)

This species has reached both coasts of North America and is always a possibility when spring storms move west across the North Atlantic. Sufficiently similar to the American species of golden-plovers to engender real problems of identification, it may be more regular than we think. A familiar bird of farmlands in Europe, it has habits much like those of its golden relatives.

Size Weight 175 g, length 25 cm (10 in), bill 2.3 cm, tarsus 4.1 cm.

Plumages Moderate plumage dimorphism. Bill black, legs dark gray. *Breeding adult* gold- and black-spangled above, with prominent white supercilium and conspicuous white line down sides separating gold of upperparts from black of underparts. Lower belly and undertail coverts white; otherwise quite variable below. Males in northern part of range typically colored rather like Pacific Golden-Plover in breeding plumage, those in southern part of range usually with less black beneath. Females in all populations with less black on underparts than males, so northern females much like southern males. Those with least black have only narrow black line up breast and foreneck connecting blackish throat with black lower breast and belly. *Nonbreeding adult* brown above, heavily spangled with gold. Underparts yellowish to white, finely marked with brown. *Juvenile* much like nonbreeding adult, more so than in other golden-plovers, and differs in plumage pattern primarily by being more heavily barred on underparts, bars often extending onto belly. Tail less regularly barred, pattern more broken up. Easily distinguished in early fall, when adults still with traces of black and worn tertials and primaries, but increasingly difficult as adults complete autumn molt.

Identification Slightly larger than both regularly occurring North American golden-plovers but so similar that identifying it necessitates careful scrutiny, especially if distinctive white underwings cannot be seen. European slightly heavier bodied than two American species, but bill not larger, so looks rather small billed compared with relatives. Wing not especially long, typically reaching tail tip, but tertials relatively short so primary projection rather long. Vagrants more likely on Atlantic coast, where confusion with American Golden-Plover unlikely when latter in full *breeding* plumage, with black sides and undertail coverts, but sexual dimorphism in both complicates identification. However, European has distinctly shorter wings not usually extending beyond tail tip. Vagrant occurrence in Alaska indicates rare overlap with Pacific Golden-Plover, much more similar in shape and in breeding plumage. In front view, European has narrowest strip of black between white neck stripes, gold-spangled feathers extending onto sides of breast; never looks as if white on neck ends at sides, and less black in white side stripe. *Nonbreeding* and *juvenile* birds brighter in European than American, recalling similar difference between Pacific and American but even more so. Winter adult European enough like juvenile plumages of all three species that confusion with adult American and Pacific unlikely. Juveniles much more similar, especially European and Pacific. European distinguished from American by overall more golden-yellow appearance as well as shorter wings, but distinguished from Pacific with difficulty. Pacific has more patterned head, with more prominent supercilium and dark ear coverts; European has plainest head among *Pluvialis*. Also, upper breast and belly somewhat plainer in Pacific than in European. Typical call of European quite different from that of Pacific. See American Golden-Plover for other shorebirds that might be mistaken for golden-plovers.

In Flight From above essentially all brown, identical to two golden-plovers usually seen in North America, but from above shows slightly more conspicuous white wing stripe and from below contrasting white rather than gray underwings. Both nonbreeding- and breeding-plumaged birds should be identifiable from underwing color, otherwise a difficult call. Lacks black axillars of Black-bellied Plover in all plumages.

Voice Flight call a plaintive double whistle that stays mostly on one level, unlike calls of other two golden-plovers which clearly change pitch. Varied descriptions indicate there may be overlap among all three species.

Behavior Much like that of other golden-plovers, and much more likely to be seen in field than on beach or mudflat. In large flocks where common but visits this hemisphere one or a few at a time.

Habitat Tundra in far north, upland moors and peatlands at lower latitudes for breeding. Grassland, pastures, and open farmland during migration and winter. Not usually in intertidal zone, but in some areas common on mudflats in autumn.

Range Breeds from northeastern Greenland and U.K. across Scandinavia to western Siberia. Winters farther south in western Europe. Migrants over fairly short distances in between. Vagrant to Atlantic Canada, almost all in spring; one winter record in southern Alaska. Winters much farther north than other golden-plovers, and any wintering golden-plover on American Atlantic coast should be carefully scrutinized.

5.1 Breeding male European Golden-Plover. Continuous white stripe on side, golden spangles extending onto breast, and relatively short bill and legs indicate species. Unbroken black underparts indicate male of northern populations most likely to occur in North America. Eastport, Newfoundland, Canada, Apr 2002 (Bruce Mactavish).

5.2 Juvenile European Golden-Plover. Strongly patterned, more so than other golden-plovers. Nonbreeding similar but underparts less heavily marked. Wing relatively short like Pacific, primary projection relatively long like American (this combination characteristic of European). Laitila, Finland, Sep 1996 (Henry Lehto).

5.3 Breeding European Golden-Plover. Only tundra plover with white underwings and axillars. Amount of black typical of males of southern breeding populations. North Yorkshire, England, Jun 2000 (Wayne Richardson).

6 American Golden-Plover

(Pluvialis dominica)

This is the most widely distributed of our golden-plovers, breeding and migrating across the entire North American continent and wintering at the other end of the American supercontinent. Large flocks can be seen in midcontinent in spring and the North Atlantic coast in fall.

Size Weight 149 g, length 23 cm (9 in), bill 2.3 cm, tarsus 4.3 cm.

Plumages Moderate plumage dimorphism. Bill black, legs gray to black. *Breeding adult* gold spangled above, much black below, with conspicuous white supercilium extending into neck stripe. Male entirely black below, female averages duller, above similar to male but with less well defined head and neck stripe and much white in black underparts; at palest extreme, much like nonbreeding adult below. Many arrive in North America on spring migration still largely in nonbreeding plumage; these may be all or mostly females. *Nonbreeding adult* with legs gray. Upperparts look irregularly mottled or virtually plain, depending on stage of wear. Fall adults that are largely pale beneath still have gold-dotted upperparts of breeding plumage. In fall, could be mistaken for juvenile at a distance, but worn state of plumage should be apparent, and almost always some black markings beneath. *Juvenile* with legs gray. Distinguished from nonbreeding adults by spotted rather than striped upperparts; both upperparts and underparts somewhat more heavily marked. Tail plain compared with that of adults, dark bars only obscurely indicated. Some heavily marked juveniles look more like adults, but heavier ventral markings should distinguish them; first-winter birds similarly distinguished. Young birds that migrate north without molting can look very different from most golden-plovers, with almost uniform brown upperparts, gray-brown breast sharply distinguished from white belly, and conspicuous white supercilium.

Identification All golden-plovers smaller and darker in all plumages than Black-bellied Plover. See very similar Pacific and European golden-plovers. Mountain Plover, Eurasian Dotterel, and Buff-breasted Sandpiper are all upland-feeding shorebirds that are plain or patterned brown above. All are smaller than golden-plovers, Buff-breasted conspicuously so, and the two plovers have darker tail tips. Nonbreeding adults and first-summer birds can be very plain and might be mistaken for larger of brown-backed, plain-breasted plovers such as Mountain, Caspian (*Charadrius asiaticus*), and Oriental (*C. veredus*) (last two not recorded in North America), but all those species are smaller, entirely plain above (even plainest golden-plovers have some paler markings in feathers of upperparts), and have darker tip and/or paler edges on tail in flight rather than faintly indicated, even barring of goldens.

In Flight From above all brown with faintly indicated wing stripes; lighter below with gray underwings. Toes do not extend beyond tail tip. In flight, Black-bellied Plover shows much more contrast than golden-plovers because of white wing stripes and whitish tail; in breeding plumage, white underwings and undertail contrast with all or partly black underparts in Black-bellied.

Voice Typical flight call a unique, abrupt *queedle* or *klee-u,* high note that breaks in middle into shorter lower note. Both golden-plovers quite vocal and can be distinguished by calls. Breeding-ground vocalizations include loud whistles in display flight, in this species an abrupt *too-lick.*

Behavior Birds spread out over feeding areas and come together in loose roosting flocks during migration. Sometimes feed in salt marshes, where they walk as much as run because of irregular surface. Commonly occur with Pacific Golden-Plover on Pacific coast.

Habitat Relatively dry tundra for breeding, grassland in winter. Migrants on grassland, coastal mudflats, and lake shores.

Range Breeds across Alaskan and Canadian Arctic, more commonly and farther south than Black-bellied. Winters mostly in southern South America, very rarely north to southeastern USA. Migrants mostly move north through eastern Mexico and central part of continent in spring. Adults head south from Maritimes and New England across Atlantic in fall, when regular in West Indies, juveniles all across continent.

6.1 Breeding male American Golden-Plover. Golden-plovers spangled gold above, black below in breeding plumage. Broad white neck stripe and entirely black sides and undertail indicate male of this species. Nome, Alaska, USA, Jun 1998 (Brian Small).

6.2 Breeding female American Golden-Plover. Scattered white feathers below indicate sex; some have even more white. Compare retained wing coverts and tertial with new ones of male. Churchill, Manitoba, Canada, Jul 1999 (Anthony Mercieca).

6.3 Nonbreeding American Golden-Plover. Plain and dull golden-plover with conspicuous white supercilium, sparsely marked underparts. Long-winged shape typical of species. Bolivia, Mar 1999 (Sam Fried/VIREO).

6.4 Juvenile American Golden-Plover. Plumage distinguished by spangled back, heavily barred underparts. Species distinguished by white supercilium, long wings and primary projection. Ventura, California, USA, Nov 1998 (Brian Small).

6.5 Juvenile American Golden-Plover. Plain brown plover with narrow white wing stripe, little other pattern. Toe tips usually not visible in flight. Saint John's, Newfoundland, Canada, Sep 1991 (Bruce Mactavish).

6.6 Adult American Golden-Plover. Gray underwing and axillars characteristic of this and Pacific Golden-Plover. Saint John's, Newfoundland, Canada, Sep 1991 (Bruce Mactavish).

7 Pacific Golden-Plover

(*Pluvialis fulva*)

A long-distance flyer, and one of the fastest, this beautiful species moves between far northern tundra and oceanic islands as well as Asian and Australian mainland shores. Mostly a peripheral species in North America, it is, however, a familiar bird of Hawaiian parks.

Size Weight 130 g, length 23 cm (9 in), bill 2.3 cm, tarsus 4.5 cm.

Plumages Moderate plumage dimorphism. Bill black, legs gray. *Breeding adult* dark brown above, heavily marked with yellow and white, much black below. Conspicuous white stripe from forehead over eyes and down each side of neck continuous with side stripe. Male solid black below from throat to upper belly; female averages duller, about like male above but with less well defined head and neck stripe and even more white in underparts than female American Golden-Plover. Those with least black much like nonbreeding adult below. Upperparts molted very early in spring, and usually in full breeding plumage in spring migration in North America. *Nonbreeding adult* gray-brown above, typically fringed with whitish or yellow. Upperparts typically look striped but may be rather plain, depending on stage of wear. Breast and sides vary from profusely striped with brown to unmarked. This plumage rarely seen in North America except in small wintering population, and autumn adults that are largely pale beneath still have gold-dotted upperparts of breeding plumage. In fall likely to be mistaken for juveniles at a distance, but almost always some black feathers on underparts. *Juvenile* distinguished from nonbreeding adult by spotted rather than striped upperparts; both upperparts and underparts somewhat more heavily marked. Tertials notched with white or yellow right to their tips in juveniles, usually continuous pale fringe around tips in adults. Tail plain compared with that of adults, with dark bars only obscurely indicated and pale yellow or whitish notching on outer edges. Retains juvenile plumage without molting while in North America, even well into November. Some with yellow-edged feathers look more like adults, but abundant ventral markings, especially bars, should distinguish them. Immatures distinguished from nonbreeding adults by dark, worn tertials and, at least in early winter, mottled breast.

Identification Easily confused with American Golden-Plover, with which it occurs in Alaska on breeding grounds and along Pacific coast in migration. When individuals cannot be determined by plumage, proportions important in identification, and species often distinguishable by this in mixed flocks. American has relatively long wings that project well beyond tail tip, giving it more pointed look behind, whereas wings of Pacific usually just reach tail tip or slightly beyond it. Primary projection longer in American, typically with three and a half or four primaries visible beyond longest tertial, whereas in Pacific about three are visible. Difference in part because of length of tertials, which fall well short of tail tip in American and usually reach or exceed it in Pacific. Tertials vs. tail may be best mark, as other proportions definitely overlap. Also, beware birds in worn plumage that may have tertial tips worn down or birds in molt that may be missing longest tertial (or longest primaries!). Bill and legs slightly longer in Pacific for its bulk, might be helpful in direct comparison but not easily assessed otherwise. In *breeding* plumage, males easily separable by white stripe or at least scattered white feathers along sides and mostly white undertail in Pacific, entirely black sides and undertail in American (rarely, Pacific has entirely black sides and American has white feathering there, but Pacific also has narrower white neck stripe than American). Females much more similar; Americans average more black beneath, but many not separable by plumage. Male Pacific and female American could be confused, but female American with white on sides should also have it scattered elsewhere across underparts, whereas male Pacific should be solid black away from sides and undertail. In *nonbreeding* plumage, American duller, basically brown and white, Pacific with golden tones in both upper- and underparts. Same true for *juveniles*, in which American characteristically has white supercilium, Pacific yellowish; this is often good field mark but not definitive. Pacific more heavily streaked on neck and upper breast, American often more heavily barred on sides and belly. See European Golden-Plover; very similar and has occurred in Alaska.

In Flight From above all brown with faint wing stripes. Nonbreeding adults and juveniles lighter below, with gray underwings. Just like American Golden-Plover, but toes project slightly beyond tail tip.

Voice Flight call a two-noted whistle, *chuwi* (rarely *chu-tu-wee*), last note higher and accented and thus opposite of American Golden-Plover. Much more like calls of Semipalmated and Common Ringed plovers than like its near relative; also much like call of Spotted Redshank. Also a sharp, whistled *wheet* given in spring and loud, whistled *whee-ter-ee* in display flight on breeding grounds.

Behavior An upland plover much like close relatives, but spends more time on beaches than others—not surprising in a bird that visits small islands all across Pacific. Birds spread out across feeding grounds, whether stubblefield, golf course, or beach, and come together in loose roosting flocks. When feeding in salt marshes or among weedy uplands, may walk as much as run because of irregular substrate.

Habitat Somewhat moister tundra than preferred by American for breeding. Beaches and estuaries as well as grasslands during winter and migration. Migrants mostly coastal, often with American on this continent.

Range Breeds in western Alaska and west to western Siberia. Winters in Australia, Southeast Asia, and all across western Pacific from Hawaii south and west; small wintering population in coastal California. Migrants move across wide Pacific Ocean and in smaller numbers along North American Pacific coast from California to western Mexican islands, rarely inland or farther north. Probably vagrant all across Canada and USA, only a few recent records on Atlantic coast and one on Barbados.

7.1 Breeding male Pacific Golden-Plover. Like American Golden-Plover but much white on sides and undertail. Black barring in white side stripe distinguishes from European. Nome, Alaska, USA, Jun 1998 (Brian Small).

7.2 Breeding female Pacific Golden-Plover. Molt appears to be completed, so extensive white below indicates this sex. Female Pacific often shows more extensive molt of upperparts in spring than American. Midway Atoll, Hawaii, USA, Mar 2001 (Brian Small).

7.3 Nonbreeding Pacific Golden-Plover. Much more yellow apparent than in nonbreeding American, feathers of upperparts with golden edges. Wing length and primary projection typical of species. Kauai, Hawaii, USA, Mar 2001 (Brian Small).

7.4 Nonbreeding Pacific Golden-Plover. Very plain individual, as plain as plainest American Golden-Plover but more yellowish. Primary projection very short; probably wing molt not completed. Manly, Australia, Dec 2002 (Tom Tarrant).

7.5 Juvenile Pacific Golden-Plover. Profuse yellow spangles indicate juvenile golden-plover. More brightly marked than American, with all-yellowish face; less brightly marked than European. Relatively short wing and primary projection typical of species. Korsnäs, Finland, Sep 1990 (Henry Lehto).

7.6 Pacific Golden-Plover. Plain brown above with faint wing stripe, gray underwings; differs from American by longer legs, toes typically projecting beyond tail tip. Thailand, Nov 2000 (Markku Huhta-Koivisto).

Ringed Plovers *(Charadrius)*

This is the dominant group of plovers, 32 small species, 12 of them in North America (2 of them accidentally), with plain brown backs and white underparts, often with a black ring or two across the breast, white collar around the neck, and conspicuous black head markings. Several species have bright rufous breast or head markings. The greatest variation from this is seen in Eurasian Dotterel, with its patterned upperparts and brightly colored underparts. Many of the species are tropical residents, but the genus also contains high-latitude long-distance migrants.

8 Greater Sand-Plover

(Charadrius leschenaultii)

Almost *Pluvialis* in shape, this is a bulkier edition of the more widespread Lesser Sand-Plover. It is an unlikely visitor to the Americas, yet a recent record on this continent underscores the vagrant possibilities among shorebirds.

Size Weight 85 g, length 20 cm (8 in), bill 2.4 cm, tarsus 3.7 cm.

Plumages Moderate plumage dimorphism. Bill black, legs yellow-green to greenish gray. *Breeding adult* with rufous breast band. Male with contrasty black head markings. Female duller than male, usually lacking black markings on head; dullest may lack virtually all bright colors. *Nonbreeding adult* just like dullest breeding females; brown breast band complete or incomplete. *Juvenile* like nonbreeding but often browner, with scaly buff feather margins as in most ringed plovers.

Identification Large for member of ringed plover group. Smaller than Killdeer but stands as tall because of long legs. Posture tends toward horizontal, more like Black-bellied Plover, whereas Lesser Sand-Plover tends to rest slightly more upright, more like golden-plovers. Bill even longer than that of Wilson's Plover, tip looks slightly more pointed than in Lesser. Yellow-green to greenish gray legs paler than gray to black of Lesser (but some Lessers, especially young birds, have greenish legs, and Greater legs can look dark). Breeding plumage as in Lesser Sand-Plover but rufous breast band narrower, usually lacks black line behind throat (this difference only in easternmost populations of both species, those most likely to reach North America). Nonbreeding adults and juveniles like large editions of Lesser Sand-Plover, essentially identical in plumage but slightly larger and with distinctly larger bill and longer legs.

In Flight Toe tips project distinctly beyond tail, usually not in Lesser Sand-Plover, and slightly more conspicuous ("bulging" at rear) wing stripe than Lesser. Tail pattern averages different, Greater with more white on sides at base and darker bar nearer tip, making more contrasty flight pattern. Primary underwing coverts slightly darker in Lesser, showing more prominent "comma" mark at that point from below. Probably some overlap in these rather difficult characters.

Voice Flight call a short trill or rattle that may be quickly repeated once or twice, quite similar to that of Lesser Sand-Plover.

Behavior Much like that of Lesser; may take longer runs and captures larger prey.

Habitat Breeding habitat desert, semidesert, and short grasslands. Nonbreeding estuaries, mudflats, and sandy beaches.

Range Breeds from Middle East across central Asia to Lake Baikal. Winters from eastern Mediterranean and east coast of Africa east across India and Southeast Asia to northern Australia. Accidental in North America, one winter record from California; much less likely to occur than northern-breeding Lesser Sand-Plover.

8.1 Breeding male Greater Sand-Plover. Large-billed ringed plover with rufous breast band, crown, and nape. Broome, Australia, Mar 1995 (Clive Minton).

8.2 Breeding female Greater Sand-Plover. Dull copy of male; some females with very little rufous. Israel, Mar 2001 (Alain Fossé).

8.3 Nonbreeding Greater Sand-Plover. No white collar; breast band complete or incomplete. Larger than Lesser Sand-Plover, with larger bill and greenish gray legs. Perhaps immature because no sign of breeding plumage this late in spring. Broome, Australia, Apr 2003 (Clive Minton).

8.4 Juvenile Greater Sand-Plover. Not much other than large bill would distinguish this bird from Lesser Sand-Plover. Breast band present or absent. Ibaraki Prefecture, Japan, Sep 1999 (Norio Kawano).

8.5 Breeding plover wing specimens. Greater (above) slightly larger with broader white stripe on primaries. Greater Sand-Plover, Tyva, Russia, Jun 2000; Lesser Sand-Plover, Magadanskaya, Russia, Jul 1992.

8.6 Same breeding plover wing specimens from below. Greater with whiter underwings, with slightly darker contrasting lesser wing coverts, forming indistinct dark marking beyond wrist.

9 Lesser Sand-Plover

(Charadrius mongolus)

This species (formerly called Mongolian Plover), which could as well be called "Siberian Plover" from the stronghold of its distribution, undergoes a dramatic plumage change for a ringed plover. Breeding birds will be at once recognizable, whereas nonbreeding plumages may escape detection unless this species' slightly larger size among the usual small plovers calls attention to it.

Size Weight 60 g, length 18 cm (7 in), bill 1.6 cm, tarsus 3.1 cm.

Plumages Slight plumage dimorphism. Bill black, legs gray to blackish. *Breeding adult* with rufous breast. Black line on anterior edge of rufous variable in width, may be lacking. Male with black markings on head, female with these markings usually brown or rufous, but some overlap. *Nonbreeding adult* lacks all rufous and black on head and breast; has white supercilium, complete or partial brown breast band. *Juvenile* more likely than adult to have gray-green to almost yellow legs. Like nonbreeding adult but scaly effect on upperparts, feathers with fine dark submarginal line and pale buff fringe. Quite variable, either with complete brown breast band or only brown smudges on side of breast with buffy (sometimes bright) wash between them, often with front edge brown.

Subspecies Several subspecies across wide range form two subspecies groups. *Mongolus* group of northeast Asia, from which North American vagrants come, have white forehead and narrow black band in front of rufous breast in breeding plumage. Members of *atrifrons* group, which breed in central Asia, have black forehead and no black bordering breast band.

Identification Larger than common small *Charadrius* plovers (Snowy, Semipalmated, Piping) but not larger than Wilson's. Unique black head markings and rufous breast in breeding plumage very different from any other regularly occurring North American shorebird. Distinguished in other plumages from other small regular species by lack of white hindneck collar. Bill larger than that of small species but slightly shorter than large-billed Wilson's, which does not occur in parts of North America where Lesser Sand-Plover most likely to be seen. See very similar but much rarer Greater Sand-Plover.

In Flight Much like Semipalmated Plover but slightly larger, overall paler above, and with no indication of white collar. White throat and rufous breast conspicuous below in breeding plumage, much like Semipalmated and Wilson's in nonbreeding but usually without breast band. Darker primary coverts on underwing produce dusky "comma" at wrist.

Voice Flight call a short, rapid trill on one level; only plover in our region with such a call. Sounds faster and higher pitched than similar calls of Baird's and Pectoral sandpipers.

Behavior Forages in typical plover fashion, quite social where common. Often associates with other small plovers and/or sandpipers.

Habitat Breeds on dry tundra in north, alpine tundra farther south, often near water. Winters on coastal sand beaches and mudflats. Migrants mostly coastal.

Range Breeds widely across northern Asia, at high and moderate latitudes. Has bred in Alaska. Winters in Southeast Asia and Australia. Migrants mostly coastal, east regularly to western Alaska. Vagrant farther south along Pacific coast of North America to southern California, accidental farther east (Alberta, Ontario, Rhode Island, New Jersey, and Louisiana).

9.1 Breeding male Lesser Sand-Plover. Brilliantly marked and unmistakable in this plumage, with black cheeks, white throat, and rufous breast and collar. White forehead typical of eastern (*mongolus*) subspecies group, likely to occur in North America. Hyogo Prefecture, Japan, May 1996 (Norio Kawano).

9.2 Breeding female Lesser Sand-Plover. A duller version of male, still unlike any other species except Greater Sand-Plover. Smaller bill than that species, and gray legs. Sakai River, Aichi Prefecture, Japan, Apr 1986 (Tadao Shimba).

9.3 Nonbreeding Lesser Sand-Plover. Plain brown ringed plover without white collar and with or without breast band; legs usually gray. Port Hedland, Australia, Mar 1995 (Clive Minton).

9.4 Juvenile Lesser Sand-Plover. Like nonbreeding but with conspicuous pale fringes on coverts. Juveniles can have legs as greenish as those of Greater Sand-Plover; bill size best distinguishing characteristic. Tokyo, Tokyo Prefecture, Japan, Sep 2003 (Angus Wilson).

9.5 Nonbreeding Lesser Sand-Plover. Breast band and lack of collar can be seen, also ringed plover tail pattern. Wing stripe without conspicuous flare on primaries shown by Greater Sand-Plover, but note how wing position affects appearance of stripe. With Red-necked Stints. Shiokawa, Aichi Prefecture, Japan, Oct 1988 (Tokio Sugiyama).

10 Collared Plover

(*Charadrius collaris*)

This tiny and delicate plover is a truly tropical representative of its group, scattered along sandy rivers and interior wetlands as well as coastal shores. Fitted for the sand, its toes lack any indication of webbing.

Size Weight 28 g, length 14 cm (5.5 in), bill 1.5 cm, tarsus 2.5 cm.

Plumages Bill black, legs dull pale pinkish. *Adult* with black crossband on breast, often with rusty tinge (brighter in male) on crown, ear coverts, and nape (this is its "collar"). *Juvenile* lacks black on head, has incomplete brown to black breast band and faint pale cinnamon feather edges on upperparts.

Identification Smallest of American ringed plovers, with quite slender bill and relatively long legs, reminiscent of Snowy but somewhat darker and with pinkish to yellowish legs, adults with complete narrow black crossband on breast. Name notwithstanding, lacks white collar of Semipalmated, Wilson's, and Piping with which it might occur. Forehead patch larger and breast band narrower than in superficially similar Semipalmated.

In Flight Rather like Semipalmated but wing stripes slightly less conspicuous and outer tail feathers white; also lacks white collar. Flight usually low and straight.

Voice Flight call a sharp *peep* or *pit pit,* also rolling calls, mostly during breeding. Not very vocal.

Behavior Short runs and stops like all small plovers, almost ghostlike on beaches of same color. Only rarely in flocks but often associates with other plover species. Tame but fast-running when disturbed.

Habitat Usually on sandy beaches of rivers or coast.

Range Resident from Sinaloa and Tamaulipas south through much of tropical South America, including southern Lesser Antilles (resident on Grenada, casual farther north). Vagrant to USA, one spring record from southern Texas.

10.1 Adult Collared Plover. Very small, thin-billed ringed plover with characteristic head markings, no white collar. Juvenile lacks black head markings, has complete or incomplete brown breast band, faintly indicated pale fringes on upperparts. Nayarit, Mexico, Jan 2000 (Chris Wood).

10.2 Adult Collared Plover. Small plover with breast band, no white collar, inconspicuous wing stripes, and white-edged tail. Puerto Vallarta, Jalisco, Mexico, Jan 2004 (Chris Wood).

11 Snowy Plover

(*Charadrius alexandrinus*)

This species is at home on sandy coastal beaches, where it has declined greatly because of human disturbance, and inland alkaline playas, fortunately inhospitable to most human activities. On either of these pale substrates, its coloration renders it nearly invisible.

Size Weight 41 g, length 15 cm (6 in), bill 1.4 cm, tarsus 2.3 cm.

Plumages Slight plumage dimorphism. Bill black, legs gray to blackish. *Breeding adult,* in particular male, with forecrown, eye stripe, and partial breast band black; crown and nape may be orange-buff. Female with eye stripe and breast markings usually same color as back but in some with black intermixed; no buff on head. Birds in tropics may maintain this plumage during winter. *Nonbreeding adult* with no black in plumage, not distinguishable from dullest breeding females. *Juvenile* legs gray-green to greenish yellow. Like nonbreeding adult but back with scaly pattern from fine pale fringes on most feathers; quickly becomes like adult through wear.

Subspecies *C. a. tenuirostris* of Gulf coast and Caribbean very pale above, about like Piping Plover; *C. a. nivosus* of Pacific coast and interior West overall darker above, darker than Piping Plover in direct comparison. Proportion of western birds that mix with resident birds on Gulf and beyond not known. These two subspecies combined by many authors but seem recognizable. Possible vagrant *C. a. dealbatus* from eastern Asia even darker than western North American birds, almost as dark as Lesser Sand-Plover. Also larger and longer-billed than American birds, with bright buff to rufous crown and always black loral stripe (American Snowies may have touch of rufous on crown, and some have loral stripes). Rarely has almost complete breast band. In nonbreeding plumage compare with Lesser Sand-Plover as well as Snowy Plover.

Identification Readily identified among other small plovers by slender black bill, dull-colored legs, pale upperparts, and lack of breast band. Of North American species, only Piping Plover similarly pale, and it has bright orange legs and shorter bill (orange based in breeding), and some Pipings have complete breast band. See also Lesser Sand-Plover and Collared Plover. Note that white collar can be hidden when bird shortens neck.

In Flight Small, pale plover with moderately conspicuous wing stripes and darker subterminal tail bar. From below only partial breast band evident. Very similar to Sanderling in flight, but inner wing all pale, whereas Sanderling has dark lesser coverts, so fore edge of wing conspicuously dark.

Voice Flight call a short, low-pitched note, *prit* or *prit-it,* which may be extended into a trill. Territorial call a two- or three-noted whistled *pree-eet* given by males on ground, second note highest and loudest.

Behavior Among real runners in shorebird world, a pale ghost that moves rapidly across sand of same color and can be lost while in plain sight. If pushed, will fly. Forages by running and picking in typical plover fashion but occasionally probes shallowly. Foot-stirring behavior on wet substrates. Roosts on upper beach in loose small flocks of its own species, typically nestled into depressions in sand.

Habitat Sandy coastal beaches throughout year. Interior birds breed on alkaline flats associated with interior lakes, some on sandy rivers. In some areas, especially in tropical latitudes, moves out onto dark mudflats with other plovers.

Range Breeds on Pacific coast from southern Washington south to Oaxaca, in Great Basin and southern Great Plains, on Gulf coast of USA and Mexico, and in Bahamas and Greater Antilles east to northernmost Lesser Antilles. Also on west and north coasts of South America and widely in Eurasia. Winters in breeding range (coastal populations) and on both coasts of Mexico. Migrants primarily on coast, rarely through Central America. Vagrant all across North America from Alaska and Yukon to Ontario and South Carolina.

11.1 Breeding male Snowy Plover (*nivosus*). Pale back, slender bill, dull-colored legs, and incomplete breast band all indicative of species, vivid black markings indicative of sex. At bright end of variation, with narrow loral stripe (not always present), buff crown. Western subspecies. Orange County, California, USA, Mar 1998 (Brian Small).

11.2 Breeding female Snowy Plover (*nivosus*). Black head markings of male usually but not always mixed with or replaced by brown, no buff on crown. Grayland, Washington, USA, Mar 2002 (Stuart MacKay).

11.3 Nonbreeding Snowy Plover (*nivosus*). Darker head markings replaced by same color as rest of upperparts. Note dusky lores in comparison with breeding female. Late-winter feather wear obvious. Grayland, Washington, USA, Mar 2002 (Stuart MacKay).

11.4 Juvenile Snowy Plover (*nivosus*). Like nonbreeding but with feathers of upperparts conspicuously fringed. Moss Landing, California, USA, Oct 2001 (Joe Fuhrman).

11.5 Breeding male Snowy Plover (*tenuirostris*). Gulf coast subspecies like *nivosus* but averages paler, faded birds very pale. Sanibel, Florida, USA, Mar 1996 (Brian Small).

11.6 Breeding female Snowy Plover (*tenuirostris*). As in male, all plumages average paler than *nivosus*. Sanibel, Florida, USA, Mar 1996 (Brian Small).

11.7 Breeding Snowy Plover (*tenuirostris*). Small pale plover with distinct white wing stripe. Distinguished from Piping by juncolike pattern of outer tail feathers as well as smaller size. Clearwater, Florida, USA, Mar 1992 (Wayne Richardson).

11.8 Breeding male Snowy Plover (*dealbatus*). Asian subspecies, potential vagrant on North American Pacific coast. Quite distinct from American subspecies in larger size, darker coloration, wide black loral stripe, and bright buff to rufous cap (female duller). Somewhat like Lesser Sand-Plover in nonbreeding plumages but with white collar. Okinawa Prefecture, Japan, Mar 2002 (Tomokazu Tanigawa).

12 Wilson's Plover

(Charadrius wilsonia)

"Big bill" best describes this species, which uses that large appendage to capture fiddler crabs and other large prey on its beach runs. It is an exemplar of the largely resident plovers that inhabit tropical beaches worldwide.

Size Weight 68 g, length 18 cm (7 in), bill 2.1 cm, tarsus 3.0 cm.

Plumages Slight to moderate plumage dimorphism. Bill black, legs pinkish to pale orange. *Breeding adult* with conspicuous head markings and breast band, in male lores and breast band black, in female usually brown (breast band even reddish brown). *Nonbreeding adult* with black areas of breeding male replaced by brown (female little changed), and breast band may be incomplete. *Juvenile* like nonbreeding adult but above with dark submarginal lines and buff fringes, producing scaly effect. More likely to have incomplete breast band.

Subspecies West Mexican *C. w. beldingi*, most likely to occur on Pacific coast of North America, has relatively narrow forehead stripe, slightly darker upperparts, and broader breast band in comparison with Atlantic and Gulf coast *C. w. wilsonia*. South American *C. w. cinnamominus*, which breeds in southern Lesser Antilles, characterized by much rufous on head and breast band, even in males. Still confusion in literature regarding subspecies characteristics, in particular distribution of reddish coloration by sex and subspecies, and further study warranted.

Identification Most similar to Semipalmated Plover but a bit larger, with much larger bill. Darker than Snowy and Piping plovers, again with much larger bill. Lesser Sand-Plover similarly bulky but with slightly shorter bill, no white collar, and gray legs. Both species may have incomplete breast band in similar nonbreeding plumages, although this less common in Wilson's. See also Collared Plover on tropical coasts.

In Flight Much like Semipalmated Plover in pattern (brown above, conspicuous white wing stripe, white at sides of rump and tail base, and black at tail tip) but slightly larger and with distinctly larger bill.

Voice Flight call a piercing, whistled *wheet* or *whit*, sometimes repeated as a series of notes; also given as alarm call. Breeding-ground birds give rattle calls, no flight display.

Behavior This is another runner, preferring running over flying if approached, and it looks hunched and bull-necked as it runs. Usually seen singly or in small, spread-out groups, often roosting with other small plovers in loose aggregations. Prey typically large, often crabs.

Habitat Sandy beaches, some moving to mudflats in migration and winter.

Range Breeds on Atlantic and Gulf coasts from Virginia south to Belize, through much of West Indies, and on north coast of South America; on Pacific coast from Baja California south to Peru. Winters from northern Mexico, Gulf coast, and Florida south to northern South America. Migrants coastal. Vagrant north to southern Oregon, Colorado, Oklahoma, Minnesota, Ontario, and Nova Scotia.

12.1 Breeding male Wilson's Plover (*wilsonia*). Combination of large bill, complete breast band, and white collar distinctive of species, black head markings of sex. Atlantic and Caribbean subspecies. South Padre Island, Texas, USA, May 2002 (Brian Small).

12.2 Breeding female Wilson's Plover (*wilsonia*). Typically brown replaces black head markings of male. South Padre Island, Texas, USA, May 2002 (Brian Small).

12.3 Breeding female Wilson's Plover (*beldingi*). Pacific coast subspecies with white forehead narrower than in *wilsonia*, breast band averaging broader. Considerable rufous on side of head typical of female of this subspecies but may occur in others. Bahia Santa Maria, Sinaloa, Mexico, Apr 2002 (Stuart MacKay).

12.4 Nonbreeding Wilson's Plover (*wilsonia*). Much like nonbreeding female but perhaps male because of black on breast band. Breast band may be incomplete. Fort Myers, Florida, USA, Jan 1995 (Tadao Shimba).

12.5 Juvenile Wilson's Plover (*wilsonia*). Like nonbreeding but with pale fringes on feathers of upperparts (already mostly worn off in this individual). Breast band often incomplete. Fort Myers Beach, Florida, USA, Jul/Aug 2003 (Clair Postmus).

12.6 Breeding female Wilson's Plover (*wilsonia*). Like large, heavy-billed version of Semipalmated Plover but wing stripe obscure on inner wing. Clearwater, Florida, USA, May 2003 (Wayne Richardson).

13 Common Ringed Plover

(*Charadrius hiaticula*)

The Old World equivalent of Semipalmated Plover, this species is quite similar in habits and appearance, but flight calls, as usual, are definitive. The two species overlap in breeding range but retire to their respective hemispheres in winter.

Size Weight 53 g, length 18 cm (7 in), bill 1.4 cm, tarsus 2.4 cm.

Plumages Slight plumage dimorphism. Bill orange with black tip, legs orange. *Breeding adult* with conspicuous head markings and breast band; lores, ear coverts, forecrown, and breast band always black in male, more likely washed with brown in female, but some females fully black. *Nonbreeding adult* with bill mostly black, legs duller orange, all head and breast markings brown. White to pale buff supercilium more prominent, extends forward over eye. Breast band may be broken in middle. *Juvenile* with bill entirely black, legs duller orange than adult. Like nonbreeding adult but with fine dark submarginal lines and whitish fringes on feathers of upperparts conveying scaled appearance. More likely than adult to have interrupted breast band.

Subspecies North American breeders are *C. h. psammodroma*, averaging slightly larger in size than Siberian breeders *C. h. tundrae*; they cannot be distinguished in field. Birds from western Europe considered *C. h. hiaticula* and from northern Europe considered *tundrae* have breast bands wider on average than those of Siberian *tundrae* or North American *psammodroma*.

Identification Rare in North America other than in breeding range and must be carefully distinguished from very similar Semipalmated Plover. European birds larger than Siberian birds, might be distinguished from Semipalmated by size alone. Common Ringed more heavily marked, breast band averaging wider and black markings on head in *breeding* plumage more extensive. White patch on forehead usually pointed toward rear and typically meets eye, same patch in Semipalmated rounded and falling short of eye. Male Common Ringed usually has white partial supercilium behind eye, whereas male Semipalmated lacks it, but females of both have white there. Semipalmated shows more obvious narrow yellow eyering in any plumage than Common Ringed (but males of that species can show such a mark). In *nonbreeding* and *juvenile* plumages, dark lores of Common Ringed slightly wider and extend slightly lower at bill base, reaching gape, whereas in Semipalmated white of throat extends just above gape. Close look necessary to determine this but may be one of best field marks for juveniles. Suggestion in literature that juvenile Semipalmated has white feathering below posterior border of eye whereas Common Ringed does not has not been borne out by further study. Width of breast band good field mark but not definitive, as both species can widen or narrow it by neck posture. Structural differences slight but may be apparent. Common Ringed has slightly longer, more slender bill; Semipalmated bill widens more toward base. Common Ringed has less webbing between toes, none between outer two and very little between inner two; Semipalmated has conspicuous webbing between outer two and slight webbing between inner two. Also difficult to use as field mark. Voice may be best confirmation of either species where rare.

In Flight Exactly like Semipalmated Plover, notwithstanding claimed difference in wing stripes in literature.

Voice Flight call reminiscent of familiar *chuwee* of Semipalmated but more plaintive and slightly lower pitched; also less emphatic and with second syllable less sharply accented. Sounds more like *poo-eep* than *chuwee* (more of quality of Piping and Little Ringed plovers). If either species known well, other easily distinguished by different call. Breeding-ground song includes more complex rhythmic whistles and chattering sounds.

Behavior No differences observed between behavior of this species and Semipalmated Plover. Feeds on beaches, mudflats, and shores of freshwater lakes.

Habitat Sandy and gravelly beaches and other open environments, usually near fresh or salt water, for breeding. Sand beaches, mudflats, and freshwater shores during migration and winter.

Range Breeds from Baffin Island in New World to Greenland and Iceland and across northern Eurasia to Chukotski Peninsula; has bred on St. Lawrence Island, Alaska. Winters in sub-Saharan Africa and Madagascar. Migrants widely through Europe, western Asia, and Africa. Vagrant to southern Asia, Australia, and in North America, to western Alaska and far eastern Canada and New England, also Barbados; likely more common than records indicate. Common Ringed and Semipalmated clearly replace one another in each hemisphere but overlap in breeding on Baffin Island and Chukotski Peninsula; birds breeding in "wrong" hemisphere migrate to "right" one.

13.1 Breeding male Common Ringed Plover. Like Semipalmated but with broader breast band and head markings, white behind eyes, no yellow eyering. Cyprus, May 2003 (Mike Danzenbaker).

13.2 Breeding male Common Ringed Plover. This individual with narrower breast band, more like Semipalmated. No indication of webs between toes (difficult to see). Turku, Finland, Jun 1989 (Henry Lehto).

13.3 Breeding female Common Ringed Plover. Brown on ear coverts and breast band indicates sex. Turku, Finland, May 1991 (Henry Lehto).

13.4 Nonbreeding Common Ringed Plover. Bill entirely black in this plumage; no black on head, complete white supercilium. Breast band complete or, less commonly, incomplete. Cyprus, Nov 2002 (Mike Danzenbaker).

13.5 Juvenile Common Ringed Plover. Like non-breeding but with fine pale fringes and dark subterminal lines on upperparts; legs duller than in adult. Extremely like Semipalmated (see text). Aomori Prefecture, Japan, Sep 1999 (Mike Danzenbaker).

13.6 Breeding Common Ringed Plover. Basically identical to Semipalmated Plover; wing stripe may average slightly more conspicuous but not enough to be diagnostic. Flight calls distinguish species. Cleveland, England, Mar 2001 (Wayne Richardson).

14 Semipalmated Plover

(Charadrius semipalmatus)

This is the most common small plover over much of its range and is the typical representative of its group, with its single breast band and plaintive calls. Watching its feeding methods—run, stop, and run again, then pull a thin worm from the sand, or vibrate a foot on the substrate—gives insight into plover natural history.

Size Weight 47 g, length 16.5 cm (6.5 in), bill 1.3 cm, tarsus 2.4 cm.

Plumages Slight plumage dimorphism. Bill black with orange base (averaging slightly duller in female), legs dull to bright orange; narrow yellow eyering. *Breeding adult* with bright head markings and black breast band. Forecrown, loral stripe, and at least part of ear coverts black in male, more likely brown in female, but sexes overlap. Female more likely to have short white supercilium behind eye. *Nonbreeding adult* with bill black, legs dull yellow or yellow-orange. Black on head and breast band replaced by brown. White supercilium extends over and behind eye. Breast band broken in small proportion of individuals. *Juvenile* with bill and legs as nonbreeding adult. Plumage identical except usual fine black submarginal lines and pale buff fringes on feathers of upperparts causing scaly effect. Breast band more likely to be interrupted.

Identification Piping and Snowy plovers paler, often or always, respectively, lacking breast band; Snowy in addition has dull-colored legs. Lesser Sand-Plover larger and bigger-billed and lacks white collar. Common Ringed very similar; see that species. See also Little Ringed and Collared plovers.

In Flight As most common ringed plover on beaches and mudflats, archetype of group's flight pattern, brown above and white below with fairly conspicuous white wing stripes and blackish subterminal band and narrow white tip and edges on tail; sides of tail base and rump white. White collar visible from above, dark breast band from below.

Voice Flight call a clear double whistle, *chuwee,* second note emphatically pronounced and higher pitched. Only Pacific Golden-Plover among regular North American species has call similar enough to cause confusion. Breeding-ground song combines regular flight call, other whistles, and chattering trill.

Behavior On coasts often only shorebird besides Dunlin that is equally common on both estuarine mudflats and sandy beaches fronting ocean, where it runs and stops in typical plover fashion. Because common, easy to see it "foot-stirring," vibrating a forward-angled leg on substrate to disturb invertebrates at surface; their movement from this stimulus spells their doom. Most food taken right from surface but will dig down a few millimeters with short bill. Will roost with small sandpipers but usually in loosely aggregated flocks of own species. Flocks in flight usually small and spread out.

Habitat Tundra, gravel bars, and sandy beaches for breeding. Beaches and mudflats during winter. Also freshwater shores in migration.

Range Breeds across arctic and subarctic latitudes of North America, also down Pacific coast of Canada to British Columbia (very rarely farther south, farthest to interior Oregon); also at least locally on Chukotski Peninsula of Russia. Winters along coasts from southern Washington and New Jersey south through Caribbean and much of coastal South America. Migrants throughout region.

14.1 Breeding male Semipalmated Plover. Small ringed plover with white collar, conspicuous breast band, orange bill base and legs, fine yellow eyering. Black on head and breast typical of male. Nome, Alaska, USA, Jun 1998 (Brian Small).

14.2 Breeding female Semipalmated Plover. Much brown on head usually indicates female. Typical tail pattern of many ringed plovers. Distraction display at nest. Nome, Alaska, USA, Jun 1999 (Larry Sansone).

14.3 Nonbreeding Semipalmated Plover. No black on head, complete white supercilium, all-black bill (perhaps some orange because late in winter). Note web between inner and middle toe on right foot. Ventura, California, USA, Mar 1999 (Brian Small).

14.4 Juvenile Semipalmated Plover. Like nonbreeding but plumage indicated by fine fringes, legs paler than in adult. Breast band may be incomplete. Ventura, California, USA, Sep 1997 (Brian Small).

14.5 Juvenile Semipalmated Plover. Fringes typical of plumage more obscure on this individual. Note small web between inner and middle toes, more substantial one between middle and outer toes on right foot (typical of species but hard to see in field). Kent, Washington, USA, Aug 2003 (Stuart MacKay).

14.6 Breeding Semipalmated Plover. Small brown plover with conspicuous dark breast band, white collar, and white wing stripe. Tail pattern typical of ringed plovers, with darker tip and white fringe all around. Saint Marks National Wildlife Refuge, Florida, USA, May 2003 (Wayne Richardson).

15 Piping Plover

(*Charadrius melodus*)

As on the Snowy Plover, the pale upperparts of this species show its association with light-colored sand. Because of its occurrence on sandy beaches that are increasingly frequented by humans, it has declined over all of its coastal and interior range and is listed as endangered in Canada and threatened in the USA.

Size Weight 55 g, length 16.5 cm (6.5 in), bill 1.3 cm, tarsus 2.3 cm.

Plumages Slight plumage dimorphism. Bill black with orange base, legs bright yellow-orange. *Breeding adult* with complete or partial breast band contrasting with rest of plumage, often bright head markings as well. Forecrown, breast band, and fine line behind collar black in males and some females, same color as back in other females. Incomplete breast band more likely in eastern populations. *Nonbreeding adult* with bill black, legs dull yellow or yellow-orange. Like dullest breeding adult, with no black. *Juvenile* with bill and legs as in nonbreeding adult. Like nonbreeding adult but pale buff fringes on feathers of upperparts convey scaly appearance (less obvious than in darker species).

Identification Colored like very pale Semipalmated Plover, usually easily distinguishable, but identification can be tricky in some lighting conditions. Easily distinguished when comparison possible. Colored more like Snowy above but always with yellow to orange legs, stubbier bill than Snowy. Dark color of breast band in any plumage continues around to rear of white collar in Piping, not in Snowy. More different from that species in breeding plumage, with orange-based bill and more extensive black breast band.

In Flight Shows basic flight pattern of small plover—long white wing stripe and tail with white edges and extreme tip and dark subterminal band—but very pale, with conspicuous white tail base. Neck with white collar, breast with or without dark band from below.

Voice Flight call a sharp, whistled *peep-u*, unique among small plovers. On breeding ground a series of high-pitched piping calls given in flight display.

Behavior Typical plover behavior of running and pecking, also digs out shallowly buried prey; often vibrates foot. Likely to be on outermost sandy beaches rather than mudflats. Roosts with other small plovers, always in small numbers.

Habitat Sandy beaches all year, including on lakes, reservoirs, and rivers inland.

Range Breeds on Atlantic coast from Newfoundland to Virginia and on Great Plains from southern Prairie Provinces to Nebraska; also very locally south on Great Plains and around Great Lakes. Winters on Atlantic and Gulf coasts from North Carolina south to eastern Mexico, Bahamas, and Greater Antilles. Migrants primarily coastal, also scattered south of interior breeding range. Vagrant farther west, all the way to Pacific states and western Mexico, and south to Barbados.

15.1 Breeding male Piping Plover. Pale ringed plover with orange bill base and legs, complete breast band, black head markings. Sanibel, Florida, USA, Mar 1999 (Brian Small).

15.2 Breeding male Piping Plover. May have complete or incomplete breast band; only species of ringed plover group with this variation in breeding males. Long Island, New York, USA, May 2002 (Angus Wilson).

15.3 Breeding female Piping Plover. Black head and breast markings often replaced by brown in females. New Brunswick, Canada, Jun 1993 (Jim Richards).

15.4 Nonbreeding Piping Plover. Bill all black, no black on head or breast. Distinguished from Snowy Plover by shorter, thicker bill and orange legs. Sanibel, Florida, USA, Mar 1999 (Brian Small).

15.5 Juvenile Piping Plover. Like nonbreeding but poorly indicated pale fringes on coverts, further indicated by fresh tertials in summer. South Dartmouth, Massachusetts, USA, Jul 2002 (Glen Tepke).

15.6 Breeding Piping Plover. Small plover, much paler than Semipalmated with more conspicuous white wing stripe; tail pattern similar but more white at tail base. Less white in outer tail than in Snowy Plover, also shows more extensive breast band from below. Central tail feathers normally dark at tip. Key West, Florida, USA, Apr 1997 (Wayne Richardson).

16 Little Ringed Plover

(Charadrius dubius)

Its name describes this species, a more petite version of its larger ringed relatives. Common on rivers on its home continent, it is accidental on this side of the oceans.

Size Weight 37 g, length 15 cm (6 in), bill 1.3 cm, tarsus 2.5 cm.

Plumages Slight plumage dimorphism. Bill black, legs yellow. *Breeding adult* with black breast band. Ear coverts black in male, washed with brown in female, consistent sex difference. *Nonbreeding adult* with brown head and breast band. *Juvenile* like nonbreeding adult but with pale scalloped feather edgings above and may have incomplete breast band.

Identification Small plover with small head and long wings, shape reminiscent of small Killdeer as much as stockier Semipalmated and Common Ringed plovers. Bill black (orange based in similar species) and legs yellow (orange in similar species). Lower border of dark ear coverts angular, distinction from similar species. Yellow eyering even more conspicuous than in Semipalmated, although subdued in juveniles. Juvenile without conspicuous supercilium of Common Ringed and Semipalmated. Long tertials almost reach primary tips (fall well short in Common Ringed and Semipalmated).

In Flight Wings plain, lacking white wing stripe typical of most small plovers, but tail pattern much like those of related species.

Voice Flight call a plaintive but short whistled *peeo*, quite different from that of Semipalmated or Common Ringed.

Behavior Runs and stops for foraging. Usually single or in small flocks, often quite tame. More characteristic of freshwater than most other small plovers.

Habitat Open areas such as riverbanks, lake shores, or salt pans with sand or gravel for breeding, also in humanmade habitats such as gravel pits but not farmland. Nonbreeders usually on freshwater, but some move to coastal mudflats in winter.

Range Breeds across Eurasia, temperate and tropical (absent from high desert), and locally to northern Africa. Winters in tropical Africa and southern Asia. Migrants over wide front. Accidental in North America, three spring records from western Aleutians.

16.1 Breeding male Little Ringed Plover. Small, long-tailed ringed plover with slender black bill, pinkish to yellow legs, conspicuous yellow eyering, white collar. Male with head and breast markings black. Aomori Prefecture, Japan, Apr 1998 (Mike Danzenbaker).

16.2 Breeding female Little Ringed Plover. Ear coverts mostly brown, diagnostic of female. Aomori Prefecture, Japan, Jul 1999 (Mike Danzenbaker).

16.3 Nonbreeding Little Ringed Plover. No black markings on head or breast but eyering still visible in this plumage. Inconspicuous supercilium and angled ear patch good distinctions from Semipalmated and Common Ringed plovers. Cyprus, Nov 2002 (Mike Danzenbaker).

16.4 Juvenile Little Ringed Plover. Like nonbreeding but with typical juvenile scaly pattern, may have incomplete breast band. Aomori Prefecture, Japan, Sep 1999 (Mike Danzenbaker).

16.5 Little Ringed Plover. Like Common Ringed and Semipalmated in flight but with scarcely an indication of white wing stripe. Fuerteventura, Canary Islands, Dec 2003 (Göran Ekström).

17 Killdeer

(Charadrius vociferus)

Birders throughout much of North America can get an insight into shorebird breeding behavior from this species, with its kill-deering butterfly flights, its habit of nesting in places as visible as road edges, and its persistent distraction displays when its nest and young are discovered.

Size Weight 90 g, length 25 cm (10 in), bill 2.0 cm, tarsus 3.6 cm.

Plumages Bill black, legs dull pinkish. *Adult* with no seasonal plumage change, but fresh fall birds show rufous fringes on tertials and scapulars and often slightly obscured breast bands. Sexes essentially alike but claimed that males in pairs sometimes recognizable by more black on face and/or more reddish on back. Complete overlap in specimens examined. *Juvenile* with bill, legs, and plumage as in adult, but faint buff fringes visible on upperparts in fresh plumage; quickly becomes like adult through wear.

Identification Easily recognized by combination of size, plover shape and behavior, and two black breast bands. More slender and darker above than Mountain Plover, much larger and longer-tailed than smaller plovers with dark brown upperparts (Semipalmated, Wilson's). Note downy Killdeers go through stage in which they appear to have single breast band; might be mistaken for smaller species but more gangly, long tailed, and fuzzy looking.

In Flight Brown above and white below, with conspicuous white wing stripe and long, rufous-based tail, latter a good field mark. Two black bands visible from below on breast. Flight usually much more languid than in other shorebirds, but migrating birds, sometimes in flocks, can fly rapidly with rapid wing beats just like their relatives.

Voice A musical *tir-eeee* on ground, often extended with a series of single notes, *tir-eeee-dee-dee-dee-dee*, when taking off, and a harder *dee* are predator-alarm calls and/or flocking calls. Breeding-ground song onomatopoetic *kill-deer kill-deer* given by males in display flight, occasionally by females.

Behavior Perhaps most familiar shorebird to most North Americans, actually somewhat atypical shorebird, not forming big, cohesive roosting or flying flocks characteristic of so many species. Occasionally, probably during migration, a tight flock in rapid flight is seen, usually causing confusion because such sights so atypical of Killdeers. More typical is scattering of birds on watered golf course or plowed field, individual birds moving about with plover runs and stops, then flying, often noisily, to new spot. Classic "sentinel" shorebird, quick to give alarm calls warning of any foreign object intruding on its vision, and frustrating to observers trying to get closer to other species. Territorial tiffs commonplace, accelerating as breeding season approaches. Has typical shorebird flight display, "butterfly flight" with slowly beating wings and loud calls.

Habitat Open moist-grass habitats for breeding, usually but not always adjacent to wetlands, including farmland and pastures, golf courses and parks, and even gravel roofs in cities. Nonbreeders in most open habitats with short or no vegetation, including plowed fields, upper parts of beaches, and open salt marshes; mudflats only rarely used.

Range Breeds throughout southern Canada and USA, absent from high latitudes, and south over Mexican Plateau and in Greater Antilles. Winters from southern British Columbia, Utah, Oklahoma, southern Illinois, and Long Island south to northern South America, mixing with resident populations in coastal Peru and Chile. Migrants throughout.

17.1 Adult Killdeer. Unmistakable large, two-banded plover. Unusually long tail (for display?) gives it distinctive shape. Feathers around paired brood patches spread to settle on eggs. Seattle, Washington, USA, May 1986 (Dennis Paulson).

17.2 Adult Killdeer. Distraction display at nest shows striking and beautiful tail pattern. Ventura, California, USA, Apr 2000 (Larry Sansone).

17.3 Adult Killdeer. Freshly molted adult shows rich rufous feather edgings, reddish between breast bands. Kent, Washington, USA, Aug 2003 (Stuart MacKay).

17.4 Juvenile Killdeer. Faint feather fringes indicate juvenile, as does down adhering to tail tip. Texas, USA, Aug 1992 (Richard Chandler).

17.5 Killdeer. Vivid white wing stripe and long, rufous-based tail, as well as double breast bands, indicate species. Flight usually slow and languid. Grimsby, Ontario, Canada, Jul 2003 (Harold Stiver).

18 Mountain Plover

(*Charadrius montanus*)

Grassland makes good farmland, and grassland birds such as the inappropriately named Mountain Plover have suffered severely from habitat loss in the shortgrass prairies of the West. Not quite as charismatic as bison, their fate hinges on all too few preserves.

Size Weight 96 g, length 23 cm (9 in), bill 2.1 cm, tarsus 3.9 cm.

Plumages Bill black, legs grayish to pale pinkish. *Breeding adult* with forehead white, extending back as short supercilium; band across forecrown and loral stripe black. *Nonbreeding adult* lacks black head markings, pale supercilium slightly better defined than in breeding plumage. Tertials and coverts with rusty fringes that wear off later in fall; plumage becomes duller and faded during winter. *Juvenile* like nonbreeding adult but with pale buff fringes to all feathers of upperparts, faintly scalloped effect; these markings paler than rufous fringes of adults.

Identification Most like Killdeer but paler brown and lacks breast bands. Nonbreeding and juvenile Eurasian Dotterels, somewhat plain-looking upland plovers that are very rare in North America, have long white supercilium and pale line across breast. Nonbreeding and juvenile golden-plovers larger, with patterned plumage, although American can be very plain (see that species).

In Flight Brown above and white below, flight pattern relatively plain for a *Charadrius* plover, with wing stripes relatively inconspicuous, except white spots on primaries form whitish patches at midlength on wing. Dark subterminal tail bar stands out because tail slightly paler than back. Distinguished in flight from other midsized plain brown plovers by dark tail marking. No conspicuous markings below. Flight distinctive, with downcurved wings, and often low over ground.

Voice Flight call a single-noted *kip*; generally less vocal than other plovers. Breeding-ground song a rolling whistle, given in "falling-leaf" flight display.

Behavior A runner, typical of upland shorebirds on wide-open shortgrass prairies. Forages by running and stopping, spotting prey, and running again to grab it. Surprisingly, vibrates foot on dry ground to stir up prey as other plovers do on wet sand. Forms loose flocks on wintering grounds, not with other shorebird species.

Habitat Shortgrass prairie for breeding. Similar habitats and plowed fields for wintering and migration.

Range Breeds on western Great Plains from southern Saskatchewan south to New Mexico and west Texas; range much contracted from historic distribution. Winters from central and southeastern California, southeastern Arizona, and southern Texas south to northern Baja California and Zacatecas. Migrants scarce in between. Vagrant to both coasts.

18.1 Breeding Mountain Plover. Plain buffy brown plover with restricted black markings on crown and lores. Often reddish feather edges. Pawnee Grasslands, Colorado, USA, Jul 2001 (Brian Small).

18.2 Nonbreeding Mountain Plover. No hint of black on head. Pale fringes may indicate first-year bird. Fort Canby State Park, Washington, USA, Dec 2000 (Ruth Sullivan).

18.3 Juvenile Mountain Plover. Like nonbreeding but pale fringes on upperparts. Antelope Valley, California, USA, Nov 1998 (Brian Small).

18.4 Mountain Plover. Plain light brown plover with white patch on outer wing, scarcely a hint of stripe on inner wing. Tail plain but dark tipped. California, USA, Nov 1979 (Larry Sansone).

19 Eurasian Dotterel

(Charadrius morinellus)

Dotterels deviate from their genus substantially in having patterned upperparts and a dark belly in breeding season, convergent in these ways with the tundra plovers, perhaps because this coloration is appropriate for tundra breeding and aerial displays. The slightly duller males are responsible for most parental care.

Size Weight 108 g, length 23 cm (9 in), bill 1.6 cm, tarsus 3.6 cm.

Plumages Slight plumage dimorphism. Bill black, legs dull yellow to yellow-green. *Breeding adult* looks gray-brown above and dark brown below at distance but handsomely colored at close range, with white supercilium, chestnut belly, and pale breast "smile." Sexes may be equally bright, but males often duller than females, with streaked crown, less contrasty head and breast pattern (gray of foreneck more likely streaked with white), and paler belly. *Nonbreeding adult* with upperparts as in breeding adult but head and underparts less distinctly marked, duller and paler, belly pale gray to whitish. Some faint streaking on neck and breast, supercilium with more buff. *Juvenile* looks striped in comparison with adult. Fringes on upperparts paler, usually interrupted at tips, producing more broken pattern. White crossband on breast less distinct than in adult and bordered by short streaks fore and aft, belly somewhat buffier.

Identification Most similar to golden-plovers in dull plumage but distinctly smaller and smaller-billed, with longer pale supercilium, pale breast band, buff-washed belly, and yellow legs. These marks also distinguish it from Mountain Plover of same habitats.

In Flight Quite plain, with little indication of wing stripe. White corners and dark subterminal band on tail not duplicated in other plovers of similar size with plain wings. White shaft of outer primary surprisingly conspicuous against brown wing. White supercilium and pale line across brownish breast add distinctiveness, as well as rather dark gray underwings.

Voice Usual call a soft trill or twittering, also a soft, repeated single note—*put, put.*

Behavior Associates in small flocks where common; single on this continent but not usually associated with other species. Feeds in typical plover fashion.

Habitat Breeds on open tundra and grassland, both lowland and well up in mountains. Winters on short grassland and open fields, more rarely on beaches.

Range Breeds from Scotland across northern Eurasia to Chukotski Peninsula, also on steppes of central Asia. Winters very locally in North Africa and Middle East. Migrants across Europe and Asia, east rarely to western Alaska in spring (may have bred there). Vagrant in fall and winter along Pacific coast south to northern Baja California.

19.1 Breeding male Eurasian Dotterel. Small-billed, brightly marked plover with vivid white supercilium and breast "smile" and dark under-parts. Female similar but more likely to have solid gray foreneck. Arjeplog, Sweden, Jul 1988 (Mike Danzenbaker).

19.2 Nonbreeding Eurasian Dotterel. Less brightly marked and with pale underparts but still retains characteristic dotterel markings. La Misión, Baja California, Mexico, Feb 1998 (Brian Small).

19.3 Juvenile Eurasian Dotterel. Vividly striped and fringed upperparts, breast more heavily streaked than in adult, supercilium buffy. Pori, Finland, Aug 1994 (Henry Lehto).

19.4 Juvenile Eurasian Dotterel. Plain brown above in flight, like small golden-plover with no hint of wing stripe; gunmetal gray underwings and conspicuous white outer primary shaft. The Netherlands, Aug 2000 (Marten van Dijl).

Oystercatchers (Haematopodidae)

Oystercatchers are rather uniform, bulky shorebirds with long, chisel-shaped red bills and short, thick pinkish or red legs. Species that live on rocks tend to be entirely black and dark brown, those on sand and mud black or brown above and white below, with conspicuous flight patterns. All use their bills to pry limpets from rocks or open bivalve shells. Young are fed by adults, unusual in shorebirds, and can be seen begging much like young gulls. There are 11 species worldwide, 2 of them regular in North America and 1 an accidental.

20 Eurasian Oystercatcher

(Haematopus ostralegus)

This oystercatcher is even flashier than its Atlantic American counterpart, and its black back and red legs should make it noteworthy any time it appears on this continent, though its accidental status makes that unlikely. Its fascinating foraging habits are probably the best studied of any shorebird.

Size Weight 550 g, length 42 cm (16.5 in), bill 7.5 cm, tarsus 5.1 cm.

Plumages Eye red, bill and eyering scarlet, legs pinkish red. Female averages slightly larger and with longer, more slender bill than male, useful in sexing pairs. *Breeding adult* black and white, *nonbreeding adult* differs by addition of narrow white bar across throat. Male bill slightly darker and redder, female bill slightly duller and oranger, should be evident in pairs. *Juvenile* with dark-tipped bill, brown eye, grayish legs, browner back with scaly buff fringes. Tail band narrower than in adults. Bare-part colors get brighter over first year of life; immatures in plumage identical to adult recognized by dark bill tip.

Subspecies *H. o. ostralegus* of western Europe likely subspecies on Atlantic coast. *H. o. osculans* of Siberia, possible on Pacific coast, differs by slightly longer bill, slightly shorter wing stripe (virtually to tip in *ostralegus*, falls obviously short in *osculans*), dark shafts on outermost primaries.

Identification Very similar to American Oystercatcher but back and wings entirely black rather than brown, white throat band in nonbreeding plumage. Bare-part colors different, with eye and legs red, latter darker and brighter than in American.

In Flight Eurasian has more extensive white in wings than American, especially in primaries, where white stripe extends almost to end of wing. In addition, Eurasian has lower back as well as rump and tail base white, extending forward in point as in dowitchers, whereas American has brown back and white rump and tail base. This would be important for distinguishing juvenile Eurasian, browner above than adults.

Voice Calls loud, piercing whistles, much like those of North American–breeding oystercatchers but more abrupt.

Behavior Typical oystercatcher, moving slowly over mudflats, sand beaches, and rocky shores in search of bivalve mollusks, which it stabs or hammers open, and other invertebrates. Same birds may take earthworms in uplands.

Habitat Coastal birds live on sandy or gravelly shores, even salt marshes. Unlike American-breeding species, often nests and even more often feeds in open habitats well away from coasts or even away from water, including farmland. Inland nesting more common in eastern Europe and western Russia. Forms large flocks where common.

Range Breeds across Eurasia from Iceland to Kamchatka, much more local in Asia than Europe. Many birds resident, others move south to Mediterranean, North Africa, Arabian Peninsula, and east coast of Asia; interior birds all migratory. Accidental in North America, two spring records from Newfoundland; should be possible on Pacific coast.

20.1 Breeding Eurasian Oystercatcher. Easily distinguished from American Oystercatcher by all-black upperparts, red eyes and legs. Outer Hebrides, Scotland, Jun 1993 (Mike Danzenbaker).

20.2 Immature Eurasian Oystercatcher. White extending up lower back and across most of outer wing distinguishes from American. White throat band indicates nonbreeding plumage, dark eyes and dark-tipped bill indicate immaturity. Cleveland, England, Sep 2003 (Wayne Richardson).

21 American Oystercatcher

(*Haematopus palliatus*)

This boldly patterned species is a familiar inhabitant of sandy beaches from Canada to Argentina. Unlike some other pairs of patterned and black oystercatchers on other continents, it overlaps scarcely at all with Black Oystercatcher but hybridizes with it where it does so.

Size Weight male 568 g, female 638 g, length 41 cm (16 in), bill 8.8 cm, tarsus 6.0 cm.

Plumages Slight size dimorphism, females larger and longer-billed but not obvious in field. Eye pale yellow, eyering and bill bright red, legs pale washed-out pinkish. *Adult* with black head and neck, brown back, remainder white. Male bill slightly brighter red, female bill duller, more orange, difference apparent in pairs. *Juvenile* like adult but with whitish scalloped fringes on back and wing coverts, narrower bill with dark tip, brown eye, dull eyering, and grayish legs. Eye, eyering, and legs get brighter over first year.

Subspecies *H. p. palliatus* subspecies of Atlantic and Gulf coasts. *H. p. frazari* of Pacific coast often shows dark spotting on lower breast, dark markings on uppertail coverts, and less white in wings (narrower white band across secondaries and no white in primaries). However, some birds from range of *frazari* look exactly like *palliatus*. Characteristics of *frazari* in fact might reflect past history of hybridization with Black Oystercatcher.

Hybrids with Black Oystercatcher show plumage patterns in between the two species, anything from American-like with some black on belly to Black-like with some white in wings and underparts. Only on Pacific coast, mostly in Baja California but also recorded in southern California.

Identification Nothing else like this bird in North America, once oystercatcher bill seen. See accidental Eurasian Oystercatcher.

In Flight Dark bird with wide white wing stripes and tail base, long red bill. In both North American–breeding oystercatchers, wings more rounded than an typical shorebirds. Flight pattern most like Hudsonian Godwit but more striking.

Voice Flight call loud, clear, high, repeated whistles, *wheep, wheep, wheep*, accelerating into rapid piping whistles during ground and flight displays in breeding season.

Behavior Walks slowly and methodically, opening bivalves with knifelike bill. Stabs into open bivalves to cut adductor muscles, removes them from rocks, and hammers them open, or pries them out of sand. May probe in mud, even shallow water. Also eats wide variety of other invertebrates. Usually only a few together but may aggregate into large roosting flocks.

Habitat Coastal sand beaches and bays, mudflats, and oyster beds. More often on rocky coasts on Pacific side and in West Indies.

Range Resident on Atlantic and Gulf coasts north to Maine, straggling northward to Nova Scotia; also entire west coast of Mexico, rarely north into southern California. South of there throughout Caribbean and local along both coasts of Central and South America to Chile and Argentina. Vagrant inland to Ontario, once all the way to Idaho.

21.1 American Oystercatcher (*palliatus*). Unmistakable large brown, black, and white shorebird with huge red bill. No seasonal plumage change. Atlantic coast subspecies. Estero Beach, Florida, USA, Mar 1999 (Brian Small).

21.2 American Oystercatcher (*frazari*). Pacific coast subspecies virtually identical to *palliatus* but often with some black on breast. Islas San Benito, Baja California, Mexico, Feb 2003 (Bernardo Alps).

21.3 Juvenile American Oystercatcher (*palliatus*). Fringed upperparts, dark eye, and dark bill tip characteristic of plumage. Long Island, New York, USA, Aug 1995 (Tom Vezo/VIREO).

21.4 American Oystercatcher (*palliatus*). Unmistakable by red bill, dramatic flight pattern duplicated only by European Oystercatcher and Hudsonian and Black-tailed godwits. Stone Harbor, New Jersey, USA, Jun 1997 (Jerry and Sherry Liguori).

21.5 American Oystercatcher (*frazari*). In flight, this subspecies shows less white in wing stripe, usually not reaching primaries. Reduced white, as well as black on breast, could be consequence of hybridization history with Black Oystercatcher. Bahia de Los Angeles, Baja California, Mexico, Dec 1989 (Jim Rosso).

22 Black Oystercatcher

(Haematopus bachmani)

This species is perfectly camouflaged until it shows its bill, a shocking sight against dark plumage and dark substrate. It is still common on Pacific rocky shores, although the *Exxon Valdez* oil spill showed that even wilderness habitat is not impregnable to human alteration.

Size Weight 555 g, length 41 cm (16 in), bill 7.0 cm, tarsus 5.2 cm.

Plumages Slight size dimorphism, females larger with slightly longer, narrower bill, difference evident in pairs. Eye yellow, bill and eyering bright red, legs washed-out pale pinkish. *Adult* looks black; at close range apparent that only head and neck black, rest of body dark brown. Female bill slightly more orange than male's, difference might be apparent in pairs. *Juvenile* much like adult but with fine pale scallops all over upperparts; bill narrower, duller, dark tipped, eye brown, eyering inconspicuous, and legs grayish. Eye, eyering, and legs get brighter over first year, not fully colored until second year.

Identification Unique among North American shorebirds. Surprisingly hard to see on rocks but identifiable from great distance if bill visible.

In Flight All dark with long red bill, short legs not visible, wings bowed. Flight usually low over water but may fly higher in chases and when flying up to an island roost. Could only be mistaken for a crow in flight but wings held bowed down rather than flat, and bill usually visible.

Voice Exactly like American Oystercatcher—loud, clear, high, repeated whistles, *wheep, wheep, wheep,* accelerating in breeding season, especially during display flights.

Behavior Quite tame for a large shorebird, usually seen walking slowly among rocks and gravel for its molluscan prey. Pries up limpets or disables mussels by a quick stab of bill. Rarely probes in sand. Usually in pairs or small groups, aggregates into groups of up to a few dozen at roosts, occasionally more than 100 together.

Habitat Islands and mainland coast with unvegetated rock and cobbles, wandering onto jetties, gravel, and sand beaches in nonbreeding season.

Range Resident on Pacific coast from southern Alaska to Baja California; some local movements, and northern populations somewhat migratory. Vagrant to Idaho and Sonora.

22.1 Adult Black Oystercatcher. Large, dark shorebird with big bright red bill, at close range obviously brown rather than black. Nothing else like it in North America. Rosario Beach, Washington, USA, May 1995 (Dennis Paulson).

22.2 Juvenile Black Oystercatcher. Dark eyes, dark-tipped bill, and whitish legs characteristic of plumage; faint feather edgings visible at close range. These birds just fledged. Sydney, British Columbia, Canada, Jul 2002 (Leah Ramsay).

22.3 Black Oystercatcher. Unmistakable if bill seen; if not, dark, low-flying, round-winged shorebird, often flying in pairs. Pacific Grove, California, USA, Mar 1984 (Peter La Tourrette).

Stilts and Avocets (Recurvirostridae)

The ten species of stilts and avocets are large shorebirds adapted to foraging in water. Avocets have medium-length legs and a bill wide at the base and somewhat flattened for much of its length, tapering to a fine upcurved point. Stilts have very long legs and straight, needlelike bills. Spending much of their time in open water and highly social in the nonbreeding, and in some cases breeding season, they are conspicuous rather than cryptic like most other shorebirds. Their loud, strident calls and habit of mobbing predators (including humans) makes them in fact the most conspicuous of shorebirds. Their pre- and post-copulatory displays make up for a lack of the song flights that characterize many sandpipers and plovers. Three species occur in North America, two regularly and one accidentally.

23 Black-winged Stilt

(Himantopus himantopus)

There is still much debate about the number of stilt species in the world, but at least some authors consider recognizably different stilts as different species, and this species differs from Black-necked in several plumage characters, although apparently not in basic natural history. The two species are unlikely to be seen together in the wild.

Size Weight 180 g, length 33 cm (13 in), bill 6.2 cm, tarsus 12.1 cm.

Plumages Eye red, legs bubblegum pink. Males slightly larger than females. *Adult* head markings variable, from entirely white to extensively black. Not correlated with sex, although males more likely to have entirely white head. Male glossy black backed, female duller (but not brown backed as in Black-necked). *Juvenile* with gray-brown back, white rear wing edge, and paler legs.

Identification White-headed individuals unmistakable; some with almost as much black as Black-necked, but always with pale gray-brown to whitish hind neck.

In Flight Just like Black-necked Stilt; in darker-headed individuals, look for black of hindneck interrupted by pale band just before mantle.

Voice Calls, both in flight and on ground, loud repeated *kik kik kik*, much like Black-necked and similarly noisy.

Behavior No differences noted between this and Black-necked; often considered conspecific.

Habitat No differences noted between this and Black-necked.

Range Breeds across Eurasia. Winters through Africa and southern Asia, where also resident populations. Accidental in North America, two spring records from Bering Sea islands.

23.1 Breeding Black-winged Stilt. Much like Black-necked Stilt but always with some white on hindneck. Pink flush at beginning of nesting, absent remainder of year. Almost entire underwing black in stilts. Montpellier, France, May 2002 (Thomas Roger).

23.2 Black-winged Stilt. Both sexes variable in head pattern. In this pair in precopulatory display, male has all-white head and glossier plumage. Ibaraki Prefecture, Japan, May 1999 (Norio Kawano).

23.3 Juvenile Black-winged Stilt. Much duller and paler than adult, with pale fringes on upperparts. Also paler than juvenile Black-necked Stilt. Chiba Prefecture, Japan, Aug 2003 (William Hull).

24 Black-necked Stilt

(*Himantopus mexicanus*)

Stilts, as their name implies, are the longest-legged shorebirds, which gives them quite a depth range for foraging in the protected waters they inhabit. "Slenderellas" would have been a good name for the group, as they epitomize slenderness. A stilt that stands as high as an avocet weighs only half as much.

Size Weight 169 g, length 33 cm (13 in), bill 6.4 cm, tarsus 11.3 cm.

Plumages Minimal seasonal plumage change but moderate plumage dimorphism. Eye red (apparent at close range), legs bright bubblegum pink. *Breeding adult* male with entire upperparts and wings glossy black; breast with pinkish tinge. Female like male but with mostly brown back, easily seen in good light; also with paler eye than male, no gloss or pinkish breast. *Nonbreeding adult* male with back less glossy, breast lacking pinkish, legs duller; no change in female. *Juvenile* with eye brown, legs grayish pink, paler than in adult. Superficially like adult but upperparts brown rather than black and with prominent buffy scalloping. Inner primaries and secondaries white-tipped, easily visible in flight but less obvious at rest.

Identification American Avocet of same habitat much bulkier with orange or pale gray head, much white above, different foraging behavior. Only North American records of very similar Black-winged Stilt were far from range of Black-necked.

In Flight Black above and white below, with wings very pointed. Underwings black, providing vivid contrast with white belly, an unusual color pattern in shorebirds. Long red legs project far beyond tail, unlike any other shorebird type. Flight halting, almost ternlike, wing beats surprisingly slow for bird of this size. Long legs probably affect flight mechanics.

Voice Calls *kek, kek, kek, kek* loudly and persistently when disturbed, individual notes rail-like in volume and stridence. Noisy on breeding grounds, much less vocal in migration and winter. Calls of large juveniles on breeding grounds quite different, very similar to *peep* call of Long-billed Dowitcher. No special breeding song, probably because no flight display.

Behavior Feeds in water, picking tiny invertebrates from surface and near surface with delicate bill. Visual feeder, but uses touch in muddy waters. Usually well spaced out while feeding (unlike avocet), but may be present in large numbers in favorable areas such as salt ponds. Breeds at fairly high densities with little territorial aggression; aggressively and noisily mobs aerial and terrestrial predators, including humans.

Habitat Breeds in shallow marshy lakes and coastal salt ponds. Nonbreeders move to coastal estuaries also.

Range Breeds locally from Washington, Alberta, Oklahoma, and Virginia south throughout lowlands of Middle America to northern South America, also West Indies. Winters from central and southern California, Gulf coast, and southern Florida south through tropical America. Migrants anywhere south of breeding range. Vagrant north all across southern Canada and out to Bermuda.

24.1 Breeding male Black-necked Stilt. Slender, black and white, with long bubblegum pink legs. Orange County, California, USA, Apr 1995 (Brian Small).

24.2 Breeding female Black-necked Stilt. Like male but with mostly brown upperparts (mantle, scapulars, and tertials). McAllen, Texas, USA, Apr 2000 (Brian Small).

24.3 Juvenile Black-necked Stilt. Much duller plumage, heavily fringed above and with paler legs and pale-based bill. In act of regurgitating pellet of undigested prey. Bear Island Wildlife Management Area, South Carolina, USA, Aug 2003 (Sharon Brown).

24.4 Immature female Black-necked Stilt. Duller head pattern and paler legs retained into first winter. Orange County, California, USA, Oct 1993 (Brian Small).

24.5 Adult Black-necked Stilt. Long pink legs trail behind flashy black and white bird, wing beats often halting. From above, white extends up lower back. San Diego, California, USA, Feb 1991 (Larry Sansone).

25 American Avocet

(Recurvirostra americana)

The alternation of light and dark feather regions produces a striking bird, whether on the ground or in flight. As in birds in many other taxonomic groups, the only seasonal plumage change occurs in the head and neck region. The curvature of the bill is presumably an adaptation for capturing surface-dwelling crustaceans.

Size Weight 304 g, length 41 cm (16 in), bill 8.8 cm, tarsus 10.0 cm.

Plumages Slight dimorphism in bill length and shape. Male with slightly longer and somewhat straighter bill than female; sex easily distinguished in pairs, less so in single birds. Bill black, legs bright blue-gray. *Breeding adult* with head and neck orange. *Nonbreeding adult* like breeding but with gray replacing orange of head and neck. *Juvenile* much like breeding adult, including orange head and neck (often with less orange), but bill shorter at first, tertials and primaries unworn.

Identification No other large, brightly marked shorebird like this. Mostly pale above and different shape distinguish from dark-backed, short-necked, and short-legged American Oystercatcher and dark-backed, very slender, and long-legged Black-necked Stilt.

In Flight Remarkably ducklike, with bulky body, fast wing beat, and rapid flight. Mostly whitish with black and white patterned wings, black scapular lines. Tail light gray, slightly darker than lower back and mantle. Long blue-gray legs extend well behind tail. Pale body and vivid wing pattern unique among shorebirds and similar-sized water birds such as ducks.

Voice Call a repeated, piercing, musical whistled *wheet* or *pleep*, often given when flying about an intruder. Calls much higher pitched than comparable calls coming from stilts in same habitat. Birds in migration just about silent; characteristic vocalizations are predator-alarm calls rather than flocking calls. Like stilts, no special song during breeding displays.

Behavior Feeds mostly in water but also on wet mud, pecking rapidly or sweeping bill back and forth to capture insects and crustaceans. May submerge entire front part of body while feeding, and often swims. Gathers in large, dense flocks for feeding and roosting in nonbreeding season and flies in flocks of up to a few dozen. Very noisy "sentinel" species in breeding season, breeding in loose colonies and aggressively mobbing intruding predators, including humans.

Habitat Breeds in freshwater and alkaline marshy lakes, also salt marshes and salt lagoons in California. Migrants in similar habitats, also winters on coastal bays and estuaries.

Range Breeds throughout West, north to southern parts of Prairie Provinces, east to western Great Plains, and south through Mexican Plateau. Winters in California, both coastal and interior, locally along Gulf coast, and in southern Florida; also south to southern Mexico, casually farther south. Migrants anywhere in West, also all along south Atlantic and Gulf coasts, mostly from Delaware south. Vagrant north to southern Alaska and all across southern Canada to Nova Scotia and Greenland and south through much of Caribbean.

25.1 Breeding American Avocet. Upturned bill, rusty head and neck, and blue-gray legs distinctive, along with black-and-white plumage. Relatively straight bill indicates male. Ventura, California, USA, Jul 1998 (Brian Small).

25.2 Nonbreeding American Avocet. Gray replaces rusty in this plumage. Female in front, with more curved bill. Newport Beach, California, USA, Nov 1997 (Joe Fuhrman).

25.3 Juvenile American Avocet. Rusty head like breeding adult, usually not as bright; also note fringes on tertials. Salinas River mouth, California, USA, Sep 1992 (Frank Schleicher/ VIREO).

25.4 Nonbreeding American Avocet. Vivid black and white body and wings, rapid ducklike flight. San Francisco, California, USA, Sep 2001 (Wayne Richardson).

Jacanas (Jacanidae)

Jacanas are atypical shorebirds, with their rail-like bill, gallinule-like plumage, and long legs and feet with very long toes and nails. These attributes fit them admirably for a life on floating aquatic vegetation in tropical freshwater marshes. Largely resident, they have rounded wings and relatively weak flight. Jacanas have a fascinating lifestyle, with the fiercely territorial females polyandrous, defending territories on which one or more males incubate their eggs and care for their young. Sharp-pointed spurs at the wrists are used for display and fighting. Of the eight species worldwide, two occur in North America.

26 Northern Jacana

(Jacana spinosa)

Throughout much of tropical America, flashy long-legged jacanas, more like gallinules than shorebirds, can be seen striding across floating carpets of green. These colorful shorebirds occasionally extend resident populations into southern Texas but are absent from the USA in most years.

Size Male weight 82 g, length 22 cm (8.5 in), bill 1.8 cm, tarsus 5.2 cm; female weight 135 g, length 24 cm (9.5 in), bill 1.9 cm, tarsus 5.4 cm.

Plumages Great dimorphism in size, male a third smaller than female. Bill and three-lobed frontal shield yellow, bill base pale blue. Conspicuous yellow spur at bend of wing. Legs light green. *Adult* strikingly colored, chestnut all over with glossy black head, neck, and upper back and breast. *Juvenile* with frontal shield and wing spur tiny, bill dark with yellowish base, legs a bit duller green than in adult. Quite differently colored from adult, brown above with reddish feather edgings and white below; dark brown hindneck, white lores, and conspicuous buffy white supercilium. Chestnut lesser wing coverts and rump acquired before fully adult plumage.

Identification Nothing else looks like an adult Northern Jacana, if one discounts much stockier gallinules and moorhens. Latter show no chestnut in plumage. Juvenile jacanas look a bit more rail-like, perhaps most like juvenile Purple Gallinule, but vivid pale supercilium distinctive. Still quite unlike any other shorebird in appearance and behavior. See Wattled Jacana for differences between these two species, rarely found together.

In Flight Rounded wings, yellow to chartreuse flight feathers in any plumage, and long legs dangling behind allow easy identification both of all-dark adults and white-bellied juveniles. Flight straight as in other shorebirds but alternate rapid fluttering and gliding, may be prolonged over large marshes.

Voice Loud cackling calls, single or chattering, in response to each other or other animals.

Behavior Wades in shallow water and at shore but more likely to be seen walking on floating vegetation in deeper water, stepping lightly on very long toes as leaves sink slowly beneath bird's weight. Conspicuous rather than skulking, with explosive noisy takeoff when flushed. May gather in flocks in nonbreeding season.

Habitat Freshwater marshes with much floating vegetation. Birds disperse surprisingly long distances to temporary ponds during rainy season, nomadic behavior easily explaining fluctuations in Texas status.

Range Resident from Sinaloa and Tamaulipas (sporadically to southern Texas, most October–April, bred previously but not recently) south through Middle America to western Panama, and on Cuba, Jamaica, and Hispaniola. Vagrant to western Texas and southern Arizona.

26.1 Adult Northern Jacana. Colorful chestnut and black gallinule-like, long-toed shorebird with tri-lobed yellow wattles and blue at bill base. Bright yellow flight feathers characteristic. Turrialba, Costa Rica, Aug 1975 (Terrence Mace).

26.2 Juvenile Northern Jacana. Shaped like adult but very differently colored; still different from all other shorebird types. Yellow flight feathers as in adult. Isla Cozumel, Mexico, Nov 1983 (Dennis Paulson).

27 Wattled Jacana

(Jacana jacana)

This species is barely present in North America but takes over from Northern Jacana as one moves southward into South America. The two are similar in every way but appearance.

Size Male weight 100 g, length 22 cm (8.5 in), bill 1.8 cm, tarsus 5.5 cm; female weight 145 g, length 24 cm (9.5 in), bill 2.0 cm, tarsus 5.9 cm.

Plumages Great dimorphism in size, male a third smaller than female. Bill yellow, base and two-lobed frontal shield and wattle on either side of bill red; prominent spur at bend of wing yellow. Legs light green. *Adult* entirely black with chartreuse flight feathers. *Juvenile* brown above, white below, with black crown and hindneck, white supercilium. Legs may be paler, even yellowish.

Subspecies *J. j. hypomelaena* in Panama and northern Colombia, quite different from subspecies farther south, all of which look more like Northern Jacana with black head and neck and chestnut body but still differ in head and bill characters and presence of some black on belly.

Identification Although rarely found together, easily distinguished from Northern Jacana by all-black coloration and red wattles. Juveniles similar, but if rudimentary frontal shield and wattles can be seen, pinkish in this species and yellowish green in Northern. Juvenile of *hypomelaena* has black rump and black on flanks, those areas chestnut in Northern. Known overlap only on Pacific side of Panama and Costa Rica. Few apparent hybrid specimens from Veraguas, Panama, look like Wattled but show maroon on back and three-lobed frontal shield.

In Flight Exactly like Northern Jacana, distinguished by black coloration.

Voice Cackling call like Northern Jacana.

Behavior Apparently exactly same as northern relative. Female mating with male that has lost its mate kills its chicks and quickly replaces them with her own eggs.

Habitat Freshwater ponds, lakes, and marshes with floating vegetation. Colonizes newly flooded areas in wet season.

Range Resident from central Panama south through much of tropical South America to northern Peru west of Andes, northern Argentina east of Andes. Vagrant to southwest Costa Rica.

27.1 Adult Wattled Jacana. Much like Northern Jacana but red wattle very different; Panamanian subspecies *hypomelaena*, in range of this book, entirely black. Note very long toes underwater. Calabozo, Venezuela, Dec 2000 (Dennis Paulson).

27.2 Juvenile Wattled Jacana. Essentially identical to juvenile Northern Jacana but almost no overlap in range. Jacanas walk on floating vegetation. Laventille, Trinidad, Mar 1984 (Dennis Paulson).

Sandpipers (Scolopacidae)

This family includes 89 species worldwide, 51 of them regular in North America (plus 13 vagrants and 1 perhaps extinct species). Sandpipers differ from plovers in their foraging method, a continuous moving forward, whether on land or in the water, and they feed visually or tactilely. Bill and leg length vary greatly, just about to the extremes among shorebirds (stilts have longer legs), but bill length is usually longer than in plovers. All sandpipers have four toes, except Sanderling (the hind toe is often lost in bird species that run).

Tringine Sandpipers (Tringini)

Members of this group of 17 species have been called "marshpipers" and "tattletales" and are among the most elegant of shorebirds. Fifteen of them can be found in North America. Many forage in water and have long legs and a relatively long bill and neck, and they carry themselves with poise. This surely correlates with their alert view of the world, as most of them are solitary or occur in small flocks, so they must maintain a vigil against potential predators. They are "sentinel" shorebirds, among the first to call and take flight when an aerial predator such as a falcon appears. Most are pickers but able to probe if the situation demands. Not all members are elegant; the Willet is shaped like a dumpy yellowlegs, and those that feed on land are short legged. The Spotted Sandpiper's teetering behavior places it closer to the "cute" category. Most of the tringines dip or bob their foreparts up and down at irregular intervals, another characteristic of the group. It is not known whether this has to do with improved vision or is a signal calling attention to the bird.

28 Common Greenshank

(Tringa nebularia)

This species is the Old World equivalent of Greater Yellowlegs, with rather similar habits. Seen only rarely in this hemisphere, it has been well studied in Scotland, so its breeding habits are much better known than those of our familiar species.

Size Weight 180 g, length 30.5 cm (12 in), bill 5.6 cm, tarsus 6.1 cm.

Plumages Bill with gray base and black tip, legs light greenish to greenish yellow. *Breeding adult* heavily marked with black on upperparts, breast, and sides. *Nonbreeding adult* without obvious black markings, upperparts plain gray-brown, mantle feathers, scapulars, coverts, and tertials edged with dark dots and white fringes, producing scalloped effect. Breast unstreaked to sparsely streaked with black. *Juvenile* like nonbreeding adult but upperparts darker, white fringes on most feather groups give strongly striped impression, neck more heavily streaked.

Identification In all plumages paler than Greater Yellowlegs, most similar common North American species, with paler upperparts and whiter, less marked breast; foreneck and breast look distinctly pale in comparison with body. Bill in each plumage with more gray than in comparable yellowlegs plumage. Legs greenish rather than yellow. Differs in addition by appearing more striped above in juvenile plumage. Also rare in North America and superficially similar to greenshank, nonbreeding Spotted Redshank has thinner bill and bright red legs. Marsh Sandpiper conspicuously smaller (like Greater vs. Lesser yellowlegs). Rare Nordmann's Greenshank (*Tringa guttifer*), similar and possible vagrant to North America, has shorter legs and heavier, more upturned bill.

In Flight Plain gray-brown above and white below with plain wings and longish bill and legs; white of rump and tail extending up lower back in a point. Spotted Redshank (red legs) and Marsh Sandpiper (much smaller and even paler) have similar flight patterns. Dowitchers, with longer bill and much shorter legs, also show this back pattern but have narrow white stripe on rear edge of wing.

Voice Flight and alarm calls similar to those of Greater Yellowlegs, similarly loud but somewhat sharper voiced (notes with hard edges) and commonly with two- as well as three-noted calls. All given on one pitch, whereas call of Greater Yellowlegs descends slightly.

Behavior Runs around actively like Greater Yellowlegs but appears to feed more on land than that species. Often feeds by scything, like yellowlegs and avocets.

Habitat Breeds in and near wetlands in both open and forested country. Migrates and winters in most wetland habitats, including lakes, marshes, rivers, and saltwater beaches and mudflats.

Range Breeds across northern Eurasia from Scotland to eastern Siberia in boreal forest zone. Winters in sub-Saharan Africa, southern Asia, and Australia. Migrants throughout those continents. Vagrant to North American coasts, regularly east to western Bering Sea islands in spring and much more rarely in fall, two fall records in California (perhaps same bird); accidental in Atlantic Canada, in fall, winter, and spring, and West Indies.

28.1 Breeding Common Greenshank. Like Greater Yellowlegs but legs green (or dull yellowish), tail whiter in all plumages. Distinguished from yellowlegs in this plumage by scapulars black striped rather than white notched. Cyprus, May 2003 (Mike Danzenbaker).

28.2 Common Greenshank. Worn, even more contrastingly marked individual in early fall. Black stripes on scapulars quite conspicuous. Rotterdam, The Netherlands, Jul 2000 (Norman van Swelm).

28.3 Nonbreeding Common Greenshank. Very plain in this plumage, with mostly white breast. Whiter breasted than Greater Yellowlegs, greenish legs and white tail typical. Werribee, Australia, Feb 2000 (Tadao Shimba).

28.4 Juvenile Common Greenshank. Even fringes on feathers of upperparts distinctive of plumage and distinguish from Greater Yellowlegs. Breast less heavily marked than in yellowlegs. Humboldt County, California, USA, Aug 2001 (Larry Sansone).

28.5 Juvenile Common Greenshank. White edging on scapulars and coverts prominent in this view; Greater Yellowlegs always shows more spotted pattern. Salo, Finland, Aug 2001 (Henry Lehto).

28.6 Common Greenshank. Like Greater Yellowlegs in flight but conspicuous white of tail and rump extends well up back. Aomori Prefecture, Japan, Oct 1998 (Mike Danzenbaker).

29 Greater Yellowlegs

(Tringa melanoleuca)

This species can often be picked out by behavior as it feeds actively, running or pacing on long yellow legs in shallow waters of all kinds. It is one of the most prevalent of the "tattletales," and its ubiquity and long presence in many regions make it a good model for understanding shorebird plumage changes.

Size Weight 160 g, length 30.5 cm (12 in), bill 5.5 cm, tarsus 6.3 cm.

Plumages Bill black, may have gray at base; legs bright yellow, varying to yellow-orange. *Breeding adult* heavily marked with black on upperparts, breast, and sides. *Nonbreeding adult* bill base with more gray than in breeding. Much plainer at a distance; most feathers of upperparts gray-brown edged with dark dots and whitish fringes. *Juvenile* bill base with more gray on average than in adults. Large white spots on edges of most feathers produce more coarsely marked appearance on upperparts than in nonbreeding adults, without black markings of breeding adults.

Identification Most similar are Lesser Yellowlegs, of similar color, and Common Greenshank, of similar size and shape. No other midsized shorebird has long yellow or greenish legs. Less similar Willet stockier with blue-gray legs.

In Flight Plain wings and white rump and tail are characteristic, long yellow legs projecting well beyond tail tip. Lesser Yellowlegs very similar but smaller, with slightly longer legs relative to size. Wilson's Phalarope and Stilt Sandpiper both have plain wings and white rump and tail but are much smaller.

Voice A loud, ringing, three- or four-noted *tew tew tew tew* or *klee klee klee*, heard at great distances from birds flying over. Breeding-ground song a ringing *whee-oodle, whee-oodle* given in flight; often heard during migration. Only Common Greenshank has similar call; all other staccato calls of this sort from other species are shorter, lower pitched, and/or more rapidly given, or more notes in series. Greater Yellowlegs also gives series of short *kip* notes which probably function as alarm calls.

Behavior Feeds mostly in water, actively pursuing invertebrate and fish prey by darting bill into water or sweeping it from side to side ("scything"), even running with it open ("plowing"). Bobs foreparts. Single or in small groups and spread out while foraging but may roost in large, dense aggregations. Alert and noisy.

Habitat Breeds in muskeg, open and sparsely wooded wet areas in boreal forest zone. Winters on saltwater estuaries and at freshwater lakes and ponds. Migrants anywhere there is shallow open water, mostly associated with estuaries and fresh- and saltwater mudflats.

Range Breeds in fairly narrow latitudinal belt all across southern Alaska and northern halves of southern provinces of Canada. Winters along coasts north to southern British Columbia and New York and across southernmost states south through Middle America and Caribbean and much of South America, most abundant on north coast. Migrants throughout.

29.1 Breeding Greater Yellowlegs. Large tringine with long neck and long, bright yellow legs. Moderately long bill often slightly upturned (minimally so in this individual). Heavily marked with black, including on sides, in this plumage. Anahuac National Wildlife Refuge, Texas, USA, May 1999 (Brian Small).

29.2 Nonbreeding Greater Yellowlegs. Plain plumage with no black markings on back or sides; tertials gray with black-barred, white edges. San Diego, California, USA, Jan 1984 (Anthony Mercieca).

29.3 Juvenile Greater Yellowlegs. Like nonbreeding but darker above, most feathers with white notches; thus looks more spotted. Bill exceptionally pale in this individual. Padilla Bay, Washington, USA, Jul 2002 (Stuart MacKay).

29.4 Immature Greater Yellowlegs. Like nonbreeding adult but note retained dark tertials of juvenile. Tail mostly gray rather than white as in greenshank. Ventura, California, USA, Sep 1995 (Joe Fuhrman).

29.5 Greater Yellowlegs. Uniformly dark above in flight, with pale rump and tail and bright yellow legs extending behind. Tail largely gray but looks white at a distance. Fort Myers, Florida, USA, May 2001 (Wayne Richardson).

29.6 Breeding Greater Yellowlegs (left) and Lesser Yellowlegs. Greater almost twice bulk of Lesser. Austin, Texas, USA, Jul 1992 (Greg Lasley).

30 Lesser Yellowlegs

(Tringa flavipes)

Lesser in name but not in charisma, this species is a petite version of the Greater, with similar long, yellow legs, alertness, and staccato calls. Its smaller bill and the sharper notes that come from it will always distinguish it from its close relative.

Size Weight 83 g, length 27 cm (10.5 in), bill 3.6 cm, tarsus 5.1 cm.

Plumages Bill black, legs bright yellow varying to yellow-orange. *Breeding adult* heavily marked with black on upperparts, neck, and breast, sides vary from sparsely to more heavily marked. *Nonbreeding adult* much plainer at a distance; most feathers of upperparts gray-brown with alternating dark dots and white fringes along their edges. Breast gray-brown, obscurely streaked darker. *Juvenile* with large white spots on edges of dark feathers producing more coarsely marked appearance on upperparts; lacks black markings of breeding adults.

Identification Differs from Greater Yellowlegs in size, no more than half the bulk but standing two-thirds as tall. Size difference pronounced in direct comparison, but birds seen by themselves less definitive. This species has short, straight bill, no longer than head (Greater has bill longer than head, typically slightly upturned). Lesser's bill always black, sometimes with touch of gray near base of upper mandible in juveniles; that of Greater extensively gray at base in nonbreeding adults and juveniles. Breeding adult Lessers only sparsely barred on sides, nonbreeding adults and juveniles have blurrier breast streaking than Greater, and nonbreeders have plainer upperparts. Accidental Marsh Sandpiper, of similar size and shape but with longer, very slender bill, has whiter breast in all plumages and usually greenish legs. Stilt Sandpiper most similar of common species but slightly smaller, different bill shape evident with good look, and totally different feeding behavior. Yellowlegs are noisy in migration, Stilt Sandpipers usually silent.

In Flight Plain wings and white rump and tail are characteristic, long yellow legs projecting well beyond tail tip. Exactly like Greater, but toes of Lesser project slightly more beyond tail, a bit of tarsus showing (not so in Greater). In Lesser foot projection about same as bill length, distinctly less in Greater. Wilson's Phalarope and Stilt Sandpiper similar to Lesser Yellowlegs but phalarope smaller with paler back, whiter breast, and much shorter legs. Stilt slightly smaller but very similar, distinguished by good look at bill or hearing flight calls of yellowlegs.

Voice Fairly loud, usually two-noted *tu tu*, much less ringing than that of Greater. Most similar call is that of Short-billed Dowitcher, usually with three notes given faster. Agitated birds often give single, sharp, whistled calls like those given by Greater. Breeding-ground song *wheedle-oree, wheedle-oree*, easily distinguished from Greater.

Behavior Often with Greater Yellowlegs. Feeding behavior similar but a bit less active, not as likely to run with open bill. Often feeds with side-to-side scything motion. Bobs foreparts. May aggregate into much larger groups than Greater.

Habitat Breeds in muskeg, open and sparsely wooded wet areas in boreal forest zone; mostly farther north than Greater, so trees smaller. Winters mostly in saltwater estuaries but also locally at freshwater lakes and ponds. Migrants anywhere there is shallow open water, both fresh and salt.

Range Breeds in fairly narrow latitudinal belt from Alaska to Hudson Bay, averaging more northerly than Greater. Winters along southern coasts from central California and Virginia southward and across southern states; generally withdraws farther south than Greater, especially on coasts. Also south through Middle America and Caribbean sparingly to southern South America, abundant on north coast. Migrants throughout, adults more common midcontinent than coastal.

30.1 Breeding Lesser Yellowlegs. Smaller version of Greater Yellowlegs, with shorter, straight bill; usually less heavily marked in this plumage, markings very sparse on sides. Galveston, Texas, USA, Apr 2003 (Brian Small).

30.2 Nonbreeding Lesser Yellowlegs. Plain plumage, upperparts marked like Greater but breast markings more obscure. Santa Clara County, California, USA, Sep 1993 (Mike Danzenbaker).

30.3 Juvenile Lesser Yellowlegs. Notches on feathers of upperparts create white-spotted effect; breast as in nonbreeding adult. Bill black in all plumages, unlike Greater. Ventura, California, USA, Sep 1998 (Brian Small).

30.4 Juvenile Lesser Yellowlegs. First feathers of first-winter plumage appearing in mantle and scapulars. Port Mahon, Delaware, USA, Sep 1984 (Dennis Paulson).

30.5 Lesser Yellowlegs. Like Greater in flight, with unmarked back and wings and whitish rump and tail, but smaller bill, longer legs (entire toes extend beyond tail). See Stilt Sandpiper. Calgary, Alberta, Canada, Aug 1983 (Dennis Paulson).

31 Marsh Sandpiper

(Tringa stagnatilis)

This is a characteristic species of shallow open water in Eurasia and Africa, but it is unlikely to be seen in North America. The very fine bill, white breast, and long legs makes it seem a miniature stilt, and it feeds in much the same way in similar habitats.

Size Weight 74 g, length 21.5 cm (8.5 in), bill 4.0 cm, tarsus 5.2 cm.

Plumages Bill black, legs greenish. *Breeding adult* barred and spotted with black on upperparts, breast, and sides. Black marks on upperparts distinctively shaped, like little flying birds. *Nonbreeding adult* much plainer, gray-brown above and white below, lacking black markings. *Juvenile* much like nonbreeding adult but feathers of upperparts darker, fringed with white, giving scalloped effect.

Identification Much like Common Greenshank, with greenish legs, gray-brown back, and white underparts but considerably smaller, with very slender, straight bill. In all plumages, loral stripe indistinct (distinct in greenshank and yellowlegs), supercilium extends behind eye (not so in others). Juvenile generally colored similar to greenshank but center of breast unstreaked. Distinguished from Lesser Yellowlegs by greenish legs, much grayer back in adult plumages, and whiter breast in nonbreeding plumage; also longer, more slender bill. Many differences comparable to those between Greater and Lesser yellowlegs. Parallel striking, as Marsh like Lesser Yellowlegs in proportion relative to larger species, and vocalizations similarly reduced. Colored similarly to Wilson's Phalarope in nonbreeding plumage, with similarly slender bill, but longer legs and bill should be obvious. Juveniles molt quickly into first-winter plumage.

In Flight White lower back, rump, and tail distinguish it from yellowlegs, Stilt Sandpiper, and Wilson's Phalarope. Smaller than Common Greenshank and Spotted Redshank, tringines with similar flight pattern; feet project farther beyond tail tip than in greenshank.

Voice Flight call a sharp *tew*, like single-noted Lesser Yellowlegs but a bit higher pitched, also may be repeated rapidly, running together; also soft, trilled calls.

Behavior Active feeder like yellowlegs and greenshank but even more committed to foraging in shallow water, where it wades with long legs and picks up tiny invertebrates with slender bill.

Habitat Much like those preferred by yellowlegs.

Range Breeds in belt including southern boreal forest and steppe, from eastern Europe east to just beyond Lake Baikal region. Winters throughout sub-Saharan Africa, Indian subcontinent, and southeast Asia, sparingly to Australia. Accidental in North America, two fall records in western Aleutians.

31.1 Breeding Marsh Sandpiper. Like Lesser Yellowlegs in size and shape but bill longer and more slender, legs greenish. Feathers of upperparts with bars and flying-bird-shaped spots. Note raised mantle feathers as often seen in foraging Ruffs. Broome, Australia, Mar 2003 (Clive Minton).

31.2 Nonbreeding Marsh Sandpiper. Very plain in this plumage, gray with fine white fringes above and pure white below. Aomori Prefecture, Japan, Sep 1999 (Mike Danzenbaker).

31.3 Juvenile Marsh Sandpiper. Darker than nonbreeding adult, pale fringes of scapulars strongly contrasting with dark centers. Breast less heavily marked than in Common Greenshank and yellowlegs. Somewhat like Wilson's Phalarope except for much longer legs. Chiba Prefecture, Japan, Aug 2003 (William Hull).

31.4 Juvenile Marsh Sandpiper. White rump and lower back conspicuous. Tail looks white in flight but is barred, as in yellowlegs and greenshank. Note long tibia. Chiba Prefecture, Japan, Aug 2003 (William Hull).

31.5 Immature Marsh Sandpiper. Plain scapulars of nonbreeding plumage contrast with white-fringed upperparts of juvenile plumage. Aomori Prefecture, Japan, Sep 1999 (Mike Danzenbaker).

31.6 Marsh Sandpiper. Like small greenshank in flight, with slightly longer legs. Bill slender and tail, rump, and back strikingly white. Thailand, Nov 1997 (Markku Huhta-Koivisto).

32 Common Redshank

(*Tringa totanus*)

With a somewhat southerly breeding range for a sandpiper and a relatively short-distance migration, this species is an unlikely candidate for North American occurrence, and indeed it has scarcely ever been seen on this continent. It furnishes only one of the innumerable good reasons for a trip to Eurasia.

Size Weight 125 g, length 25 cm (10 in), bill 4.2 cm, tarsus 4.8 cm.

Plumages Bill black, bright red at base; legs bright red-orange. *Breeding adult* brown above and white below; western European populations vary greatly in amount of breeding plumage acquired, at greatest heavily barred and streaked with black and very dark looking. *Nonbreeding adult* with duller bill and legs. Gray-brown above with upperpart feathers finely fringed with alternating white and black dots, breast finely streaked. *Juvenile* with bill as in nonbreeding adult, legs may be even paler, to yellow. More heavily marked than nonbreeding adult, with wide buffy fringes on upperparts and strongly streaked breast.

Identification Only red-legged tringine besides Spotted Redshank; differentiated from that species by smaller size, shorter legs and neck, and shorter, thicker bill. Almost dumpy for a *Tringa*, Willet-shaped and Willet-like in plainness in nonbreeding plumage. Heavily marked breeding birds can be very dark below but always obviously streaked, unlike Spotted Redshank. Legs of some Ruffs in nonbreeding plumage sufficiently bright orange to generate confusion with redshanks, but shorter-legged and -billed than those tringines, with patterned rather than plain crown and mantle.

In Flight Pale tail and white lower back pattern like Spotted Redshank, Common Greenshank, dowitchers, and others, but rear edge of wing (secondaries and inner primaries) extensively white, a striking pattern.

Voice Flight call a musical *teu*, slurring downward, sometimes extended into two notes. Alarm call a loud repetitive *teup teup teup teup*, etc.

Behavior Feeds by pecking at substrate, more rarely probing, also much time in water, where may sweep bill back and forth; often takes fish. Feeding birds spread out, flock to fly and roost, in large numbers where common. Noisy and often wary.

Habitat Open wetlands, including wet meadows, marshes, moors, and salt pans for breeding. Mostly mudflats and shallow lagoons of coastal estuaries for nonbreeding.

Range Breeds across Eurasia. Winters in Africa and Asia. Vagrant to North America, multiple birds in one spring and another in late winter in Newfoundland.

32.1 Breeding Common Redshank. Red-legged tringine liberally marked with black streaks and bars in this plumage. Heavily marked individual; many are less so. Red at bill base best developed in this plumage. Outer Hebrides, Scotland, Jun 1993 (Mike Danzenbaker).

32.2 Nonbreeding Common Redshank. Very plain plumage, with fine fringes and dots on tertials and some scapulars. Plumage reminiscent of nonbreeding Willet but legs and bill leave no doubt of identity. Llanfairfechan, Wales, Mar 2002 (John Dempsey).

32.3 Juvenile Common Redshank. Darker above than nonbreeding, with contrasting pale fringes and heavily striped breast. Bill may lack red, legs duller than adult. Turku, Finland, Jul 1987 (Henry Lehto).

32.4 Juvenile Common Redshank. Much less heavily marked than Finnish juvenile but still distinct from adult and any other species. Aomori Prefecture, Japan, Sep 1999 (Mike Danzenbaker).

32.5 Common Redshank. Only tringine with much of secondaries white, as well as tail, rump, and lower back. No other shorebird shares this pattern. Cleveland, England, Sep 2003 (Wayne Richardson).

33 Spotted Redshank

(Tringa erythropus)

The spectacular plumage change of this species is unique among tringines, and its dark juvenile plumage supports the hypothesis of juvenile shorebirds mimicking adults of their species on the breeding grounds to avoid predation. The bill is very slender and, unlike in other tringines, slightly drooped at the tip.

Size Weight 158 g, length 30 cm (12 in), bill 5.9 cm, tarsus 5.7 cm.

Plumages Slight plumage dimorphism. *Breeding adult* with bill black, base of lower mandible red, legs dark red to blackish. Male dark brown above and almost entirely black below, with white-dotted upperparts. Female averages paler on head and breast, with more white markings on belly. *Nonbreeding adult* with legs bright red-orange to dull red, much brighter than during breeding season. Gray above with pale gray wash on breast, rest of underparts white. *Juvenile* with legs as in nonbreeding adult. Plumage much darker than nonbreeding adult, brown above with white-dotted scapulars, tertials, and coverts. Underparts brownish gray, lightly to heavily and finely barred with darker gray. Darkest individuals reminiscent of breeding-plumaged adults, but pale supercilium distinctive. Tail bars slightly more conspicuous than those of adults. Rapid autumn molt to first-winter plumage, distinguished by combination of gray mantle and scapulars, strongly dotted tertials, and entirely white underparts.

Identification Unique in breeding plumage and just about as striking in juvenile plumage, very dark with long red legs. In nonbreeding plumage, most like Greater Yellowlegs of common North American shorebirds, but paler back and very white breast (more like Common Greenshank) and, of course, red-orange legs. Only other red-legged tringine is Common Redshank, but note that yellowlegs legs can approach this color when coated with material from some habitats, and other field marks have to be noted to distinguish Spotted Redshank, for example, slender and slightly droopy red-based bill.

In Flight Gray-brown above and white below, with plain wings; white of rump and tail extends up lower back in point. In breeding plumage very dark, can be almost black all over, but tail and lower back still contrasty pale. From below, white underwings in stark contrast with dark underparts. White up back distinguishes readily from yellowlegs (but consider dowitchers), but must be distinguished from similarly rare Common Greenshank and Marsh Sandpiper. Flight faster and more maneuverable than in most other *Tringa*.

Voice Flight call a loud, double whistle, *tew-weet*, more like Semipalmated Plover than repeated calls of a yellowlegs. Alarm call a rapidly repeated *chi chi chi chi*, etc.

Behavior Feeds actively by picking or jabbing, occasionally probing; in water often uses scything, regularly swimming and tipping up like duck. May feed in water in dense flocks, apparently cooperatively.

Habitat Tundra and northern forest for breeding, usually in open but near trees. Migrants and wintering birds primarily on shallow freshwater bodies but also on coastal lagoons, salt marshes, salt pans, and even mudflats.

Range Breeds in Arctic from Scandinavia east through Siberia. Winters in Middle East, Africa north of equator, India, and Southeast Asia. Migrants anywhere in between. Vagrant to both coasts of North America and less often all across interior, almost throughout year; also to Barbados.

33.1 Breeding male Spotted Redshank. Spectacular black tringine with white dots above. Bill and legs longer than in Common Redshank. Boreal forest sandpipers often perch on small branches to survey their territories. Northern Norway, Jul 1988 (Mike Danzenbaker).

33.2 Breeding female Spotted Redshank. Like male but extensively marked with white. Base of lower mandible red in all plumages. Tatsuta, Aichi Prefecture, Japan, May 2003 (Tadao Shimba).

33.3 Nonbreeding Spotted Redshank. Quite plain, pale gray above and white below. Legs brighter and paler red-orange than in breeding. Isshiki, Aichi Prefecture, Japan, Apr 1993 (Tadao Shimba).

33.4 Juvenile Spotted Redshank. Heavily marked, at darkest reminiscent of breeding adult but more extensively white above and below. Note slight droop at bill tip, unique among *Tringa*. Cyprus, Sep 2000 (Mike Danzenbaker).

33.5 Spotted Redshank. Plain wings, pale tail, and white triangle up back as in several other *Tringa*, but only one with this flight pattern and long red legs. Long bill with red base also should be visible. Fuerteventura, Canary Islands, Dec 2003 (Göran Ekström).

33.6 Spotted Redshank. White underwings stand out against dark body in breeding adults and juveniles. Legs quite long, entire toes extend beyond tail tip. Isshiki, Aichi Prefecture, Japan, Apr 2003 (Tadao Shimba).

34 Wood Sandpiper

(*Tringa glareola*)

This species seems between Lesser Yellowlegs and Solitary Sandpiper and might be confused with either one, less so in flight than on the ground. Fortunately, as is so often the case with shorebirds, its vocalizations are distinctive. Just as fortunately, *Tringa* sandpipers are nearly always vocal.

Size Weight 61 g, length 21 cm (8 in), bill 2.9 cm, tarsus 3.8 cm.

Plumages Bill black, grading to grayish to olive at base; legs dull yellow to greenish yellow. *Breeding adult* brown, feathers with black centers and white tips or partial fringes, giving spangled effect. White supercilium best developed before eye. Neck and breast prominently streaked. *Nonbreeding adult* less heavily marked above, feathers fringed with alternating black and white dots. Breast pale brown with faintly indicated streaks. White supercilium extends before and behind eye. *Juvenile* much like nonbreeding adult but pattern on upperparts quite even, white fringes on mantle and scapulars and notches on coverts and tertials. Breast finely streaked with gray-brown but more prominently so than in nonbreeding adult. White supercilium well marked.

Identification Colored somewhat like Lesser Yellowlegs, shaped more like Solitary Sandpiper. Legs somewhat in-between in both length and color so mistaken identity could go either way, but body color distinctly more like yellowlegs, with larger pale markings above. Shorter legged than yellowlegs, with more prominent supercilium (typically extending behind eye and giving capped effect, not so in yellowlegs), shorter, gray-based bill (entirely black in yellowlegs), and shorter wings. Wings reach tail tip in Wood Sandpiper, extend beyond it in Lesser Yellowlegs; primary projection much shorter than bill in Wood Sandpiper, almost as long as bill in Lesser Yellowlegs. See Green Sandpiper, casual in North America.

In Flight Brown above and white below, with plain wings and white rump and tail. Much like smaller version of Lesser Yellowlegs but legs distinctly shorter, foot projection half length of toes (all of toes and a bit of tarsus in yellowlegs). Light wing linings and overall paler color distinguish from Solitary and Green sandpipers in flight, also less likely to "tower" than those two species, but noteworthy for quick acceleration on flushing.

Voice Flight call an emphatic series of high-pitched notes, usually three or more—*pip pip pip*—sounding something like flight call of Long-billed Dowitcher but a bit lower-pitched. Quite distinct from yellowlegs or Solitary Sandpiper call.

Behavior Feeds in shallow water, wading about and picking prey from surface; not as active as yellowlegs but bobs foreparts similarly; also teeters like Spotted Sandpiper at times. Where common, may occur in large flocks; expected to be solitary in North America.

Habitat Breeds in boreal forest and forest-steppe interface, usually near water. Winters on and migrates through freshwater wetlands of all kinds, rarely in protected coastal waters.

Range Breeds across northern Eurasia, rarely east to western Aleutian Islands. Winters in sub-Saharan Africa, Southeast Asia, and northern Australasia. Migrants throughout Old World, regularly to western Alaska. Vagrant to other parts of North America, including Yukon, British Columbia, Newfoundland, and New York and few records from Barbados.

34.1 Breeding Wood Sandpiper. Like small Lesser Yellowlegs with shorter bill. Heavily patterned on upperparts and breast. Wings and primary projections shorter than in yellowlegs, good field mark. Legs yellowish or greenish. Cyprus, May 2003 (Mike Danzenbaker).

34.2 Nonbreeding Wood Sandpiper. Most patterning obscure in this plumage. Darker than Lesser Yellowlegs with more prominent supercilium. Al Ansab Lagoons, Oman, Oct 1997 (Hanne and Jens Eriksen/VIREO).

34.3 Juvenile Wood Sandpiper. Strongly patterned above, moderately so on breast, distinctly more so than nonbreeding adult. Supercilium distinguishes from Green and Solitary sandpipers. Paimio, Finland, Aug 1990 (Henry Lehto).

34.4 Wood Sandpiper. Plain back and wings, white rump and tail. Smaller and shorter-legged than Lesser Yellowlegs. Salo, Finland, May 1991 (Henry Lehto).

35 Green Sandpiper

(*Tringa ochropus*)

This Eurasian species shares with Solitary Sandpiper the habit of placing its eggs in old passerine nests in a tree. Rare as it is in North America and recorded only from the far-flung islands of the Bering Sea, in that region it is more likely to be seen than the Solitary.

Size Weight 80 g, length 21.5 cm (8.5 in), bill 3.5 cm, tarsus 3.5 cm.

Plumages Bill black with gray base, legs greenish. *Breeding adult* finely dotted above with white to buff, breast heavily streaked, sides of breast barred. *Nonbreeding adult* with dots on upperparts reduced but some may be present; face, foreneck, and breast paler, no markings on sides. *Juvenile* more heavily and evenly marked above with buff dots than breeding adult, but breast streaking less distinct, no markings on sides.

Identification Much like Solitary Sandpiper, with dark back and breast contrasting with white belly, white eyering, dark bill with slightly paler, grayish base, and greenish legs, but a bit larger and chunkier and slightly darker. Less attenuated toward rear, with shorter wing and primary projections. White supercilium from bill to eye conspicuous on dark head, often more so than in Solitary, and eyering slightly less conspicuous. Juvenile breast streaks more distinct than in Solitary, which has blurry streaks in that plumage. Solitary underwing coverts finely barred with white, not so in Green (both may hold wings up on landing). Dark breast in all plumages good distinction from yellowlegs and Wood Sandpiper.

In Flight Unmistakable because of very dark color and dark underwings (distinction from Wood Sandpiper and yellowlegs), apparently white rump, and darker tail. Most tail feathers broadly barred with dark brown, outermost feathers almost entirely white, unlike Solitary (which has dark rump and line down tail, barred outer feathers). Shorter legs than Wood Sandpiper, toe tips barely showing beyond tail.

Voice Flight call a series of high whistles much like that of Solitary Sandpiper but commonly high and low or liquid notes mixed—*weet-oo-weet* or *weet-ooeet-weet*.

Behavior Similar to Solitary, flushing from small water bodies and ascending rapidly with loud calls. Typically feeds by picking at water surface or submerging head. Often teeters like Spotted Sandpiper.

Habitat Boreal forest for breeding. In migration and winter tends to occur in all kinds of freshwater wetlands but slightly more likely than Solitary to be on saltwater mudflats and lagoons.

Range Breeds across northern Eurasia. Winters in Africa and Asia. Vagrant to North America, one fall and few spring records from Bering Sea islands.

35.1 Breeding Green Sandpiper. Very dark tringine with scattered large white dots above, heavily streaked breast. Slightly larger than Solitary Sandpiper, with shorter wings and quite short primary projection. Hisai, Mie Prefecture, Japan, Aug 2002 (Koichi Tada).

35.2 Nonbreeding Green Sandpiper. Like breeding but virtually no dots above. Characteristic tail visible, white with broad black bars. Cyprus, Aug 2002 (Mike Danzenbaker).

35.3 Juvenile Green Sandpiper. Finely and evenly spangled with buff dots above. Breast more strongly streaked than in juvenile Solitary Sandpiper. Salo, Finland, Jul 1990 (Henry Lehto).

35.4 Breeding Green Sandpiper. Mostly white tail distinguishes from very similar Solitary Sandpiper, dark underwings from Wood Sandpiper. Also tail with bars concentrated more toward tip, legs shorter than Wood Sandpiper. Nonflocking. East Yorkshire, England, Apr 1990 (Wayne Richardson).

36 Solitary Sandpiper

(*Tringa solitaria*)

One of the most appropriately named shorebirds, this species is characteristically by itself at the margins of freshwater ponds and lakes, in both the breeding and non-breeding seasons. A typical tringine in its actions, it is an especially dark sandpiper, with a dark brown back in all plumages.

Size Weight 48 g, length 20 cm (8 in), bill 3.0 cm, tarsus 3.1 cm.

Plumages Bill gray at base, grading to black at tip; legs dull greenish. *Breeding adult* with head, neck, and breast distinctly striped, stripes finer on foreneck and breast. Upperparts with scattered whitish dots. *Nonbreeding adult* plainer than breeding adult, with head/breast striping obscure, few or no pale dots above. *Juvenile* with quite plain brown head and breast, no trace of striping of adult plumages. Fine pale dots all over upperparts, more conspicuous than worn breeding adults at same season.

Subspecies Unusual in having northerly (*T. s. cinnamomea*) and southerly (*T. s. solitaria*) breeding subspecies, with poorly known but overlapping winter ranges but somewhat separate migratory routes in North America. *T. s. cinnamomea* breeds from Alaska east to Hudson Bay and is more common in migration in West. It is slightly larger with slightly paler, more olive-brown upperparts in breeding plumage, finely spotted instead of all-dark lores, and buff dots instead of whitish dots on upperparts in juvenile plumage. *T. s. solitaria* breeds from interior British Columbia to Labrador and appears to be more common subspecies in migration in East. Juveniles in migration identifiable in hand and should be in nature.

Identification Smaller and darker than Lesser Yellowlegs, with shorter neck and shorter, greenish legs. Somewhat like Spotted Sandpiper in size and habitat but darker, with dark breast and different habits; Solitary with longer legs, foraging more in water, and rarely teetering (it bobs frequently, however). Wood Sandpiper similar in size but with slightly longer, yellowish legs, generally paler than Solitary and marked with larger pale spots above; best distinction conspicuous white supercilium. Most similar species is casual Green Sandpiper.

In Flight Generally dark appearance, entirely dark underwings, tail dark with outer feathers extensively white. Nothing much like it except similarly dark Green Sandpiper, which has all-white tail. When flushed, Solitaries often fly straight up and away, perhaps an adaptation to wetlands with walls of trees around them.

Voice Loud, high, whistled *weet weet weet*; like most solitary shorebirds, usually calls when flushed. Reminiscent of Spotted Sandpiper but given on even pitch; Spotted's call varies and often drops toward end. Breeding-ground song also consist of short, high-pitched whistles.

Behavior Normally solitary as name implies, but very small flocks occasionally seen. Feeds singly at shore or in shallow water, moving along rapidly and picking invertebrates from or below surface. Bobs foreparts frequently, sometimes teeters rear as well. Flushes readily and noisily and often flies high into air ("towers") like snipes. When landing, holds wings up for a second.

Habitat Breeds on wooded lakes and ponds. Winters and migrates mostly in same habitat but more likely to be seen at open freshwater ponds and rivers, rarely on salt water.

Range Breeds across boreal-forest belt of Alaska and Canada from coast to coast. Winters throughout New World tropics, north through much of Mexico and West Indies; rarely north to southern USA. Migrants throughout but much less common west of Rocky Mountains.

36.1 Breeding Solitary Sandpiper (*solitaria*). Small dark tringine with greenish legs, conspicuous eyering. Heavily and often irregularly dotted with white in this plumage. Southern-breeding, eastern-migrating subspecies. Anahuac National Wildlife Refuge, Texas, USA, May 1999 (Brian Small).

36.2 Breeding Solitary Sandpiper (*cinnamomea*). Northern-breeding, western-migrating subspecies like *solitaria* but a bit more grayish olive, loral stripe less distinct. Seattle, Washington, USA, May 2003 (Stuart MacKay).

36.3 Breeding Solitary Sandpiper (*cinnamomea*). Dark central and barred outer tail feathers diagnostic. Streaked lores of western migrant subspecies. Seattle, Washington, USA, May 2003 (Stuart MacKay).

36.4 Nonbreeding Solitary Sandpiper (*solitaria*). Like breeding but only sparsely dotted. Santa Ana National Wildlife Refuge, Texas, USA, Jan 1989 (Mike Danzenbaker).

36.5 Juvenile Solitary Sandpiper (*solitaria*). Evenly and finely dotted with white above, breast more obscurely streaked than in adults. Jamaica Bay, New York, USA, Sep 1998 (Arthur Morris/VIREO).

36.6 Juvenile Solitary Sandpiper (*cinnamomea*). Dots on upperparts buff in this subspecies. Ventura, California, USA, Oct 1995 (Brian Small).

36.7 Breeding Solitary Sandpiper (*solitaria*). Entirely dark on upperparts and underwings; tail mostly dark with barred white edges. Often flies straight up and away when flushed. Nonflocking. Merritt Island, Florida, USA, May 1998 (Wayne Richardson).

37 Willet

(*Catoptrophorus semipalmatus*)

This unmistakable species with flashy wings and ringing song is conspicuous wherever it breeds, surprisingly far south for a sandpiper and well adapted to both coastal and interior habitats. Western and eastern subspecies pose identification challenges.

Size Weight 250 g, length 34 cm (13.5 in), bill 5.8 cm, tarsus 6.0 cm.

Plumages Bill black with gray or brownish base; legs blue-gray, sometimes with brownish tinge. *Breeding adult* with head and neck heavily but finely striped with darker brown, breast and sides barred with dark brown, often with warmer brown suffusion. Scapulars, coverts, and tertials variably streaked and barred with blackish. Intensity and distribution of markings vary greatly. Tail either unmarked or obscurely to fairly distinctly barred with darker gray-brown. *Nonbreeding adult* plain gray and white, without dark markings of breeding adult. *Juvenile* like nonbreeding adult but distinctly browner, with buff to whitish fringes all over upperparts on more lightly marked individuals. Very variable, darkest individuals brown rather than gray-brown, with scapulars and tertials complexly marked with buff notches, dark shaft streaks, and subterminal brown fringes.

Subspecies Atlantic and Gulf coast breeders, *C. s. semipalmatus,* average slightly smaller (about 50 g lighter, bill 0.5 cm shorter, tarsus 0.8 cm shorter) and darker above in breeding plumage than those of western interior, *C. s. inornatus*. Many western birds with lightly marked back and bars on undersurface on buff background, although some like eastern birds with more heavily marked back and contrasty bars on white background. Unfortunately, entire range of variation appears to occur in both populations. Much overlap in range when some western birds move in fall migration to Atlantic coast as far north as New England and winter in southeastern USA and Middle America. Has been claimed that nonbreeding *semipalmatus* a bit browner, and subspecies distinguishable in mixed winter flocks, but not well substantiated. Also, claimed differences in tail pattern do not hold up. In flight, western birds average wider white wing stripe, but overlap.

Identification Might be mistaken for Greater Yellowlegs but twice weight of that species with similar-length bill and gray legs, more methodical feeding. Much grayer than other large shorebirds (curlews, Marbled Godwit) with which it often associates in roosting flocks, but spring birds and some juvenile Willets are brownish. In nonbreeding plumage, very plain with no feather markings at all, thus colored more like nonbreeding Hudsonian Godwit but with shorter, all-dark bill and gray legs.

In Flight Unmistakable large sandpiper, light gray-brown above and white below with vividly black-and-white marked wings. Wing stripe common in many shorebirds evolved to spectacular display in this one, wide and entirely bordered by black on outer wing. Tail whitish with slightly darker tip, quite different from ringed tail of Hudsonian Godwit.

Voice Not very vocal in migration but quite noisy on breeding grounds. Breeding-ground song, usually given in flight, a ringing *pill will willet*, higher and faster in eastern breeders. Two different alarm calls frequently heard, fairly musical *kay-whuh* and, at higher intensity, series of loud *kip kip kip* calls, in some cases each note breaking in middle. Last also given as alarm and perhaps flocking call during migration and winter.

Behavior Feeds by itself by both picking and probing along shore and out into water as far as its legs will take it. At times feeds like Greater Yellowlegs, running with bill in water and capturing fish. Diet diverse, including many large items. Assembles into large roosting flocks, often with Black-bellied Plovers or larger species. Sometimes bobs like yellowlegs.

Habitat Breeds in freshwater and alkaline marshes in western interior, salt marshes on coast. Nonbreeders on coastal beaches, estuaries, and bays. In migration on lakes.

Range Breeds in Great Basin and northern Great Plains, also Atlantic and Gulf coasts from Canadian Maritimes to Tamaulipas and throughout Greater Antilles. Winters on Pacific coast from northern California (locally to southern Washington) south, Atlantic coast from Virginia south through Middle America and West Indies to Ecuador and northern Brazil, rarely farther south. Migrants possible anywhere, mostly coastal outside western USA. Vagrant north to northern British Columbia, Hudson Bay, and Newfoundland.

37.1 Breeding Willet (*semipalmatus*). Large, heavy-bodied tringine with straight bill and blue-gray legs. Heavily patterned in this plumage, especially Atlantic and Caribbean subspecies. Note toe webbing that gives it its specific name. Jamaica Bay, New York, USA, May 2002 (Angus Wilson).

37.2 Breeding Willet (*inornatus*). Western subspecies often less heavily patterned than eastern in this plumage. Benton Lakes National Wildlife Refuge, Montana, USA, Jun 1994 (Anthony Mercieca).

37.3 Nonbreeding Willet (*inornatus*). Plain gray in nonbreeding plumage. Few retained juvenile coverts and tertials visible here. Monterey, California, USA, Oct 1997 (Brian Small).

37.4 Nonbreeding Willet (*semipalmatus*). Eastern Willets can be somewhat more brownish tinged than western ones in this plumage. Very worn feathers probably indicate first-year bird. Edisto Beach, South Carolina, USA, Mar 2003 (Sharon Brown).

37.5 Juvenile Willet (*semipalmatus*). Fine fringes and notches on upperparts characteristic of this plumage. Rodanthe, North Carolina, USA, Aug 2002 (Glen Tepke).

37.6 Juvenile Willet (*inornatus*). More heavily marked juvenile, with distinct notches all along scapulars and tertials; some individuals even more heavily patterned in similar fashion. Ventura, California, USA, Aug 2002 (Larry Sansone).

37.7 Breeding Willet (*inornatus*). Unmistakable in flight, with broad white wing stripe bordered by entirely black outer wing, pale tail. Santa Clara County, California, USA, Apr 2003 (Mike Danzenbaker).

38 Wandering Tattler

(*Heteroscelus incanus*)

Speaking out seems characteristic of tattlers, thus the name. "Wandering" is appropriate as well, as this species occurs throughout the wide Pacific Ocean as a migrant. Common on rocky shores up and down the Pacific coast, it often shuns the flocks that comprise the other small rock shorebirds.

Size Weight 110 g, length 28 cm (11 in), bill 3.8 cm, tarsus 3.5 cm.

Plumages Bill black, base of lower mandible gray; legs yellow to greenish yellow. *Breeding adult* with underparts heavily barred with gray; bars usually extend to belly and undertail coverts. White supercilium moderately distinct, mostly before eye. *Nonbreeding adult* differs in being entirely unbarred beneath; breast and sides same color as upperparts, belly white. Loral stripe more distinct than in breeding adults. *Juvenile* with gray on bill base often more extensive than in adults. Differs from nonbreeding adult in having indistinct pale fringes on scapulars and tertials, often accompanied by alternating paler and darker dots, and a network of fine, pale bars on breast, visible only at very close range.

Identification Only very similar species is closely related Gray-tailed Tattler. Nonbreeding Surfbird similarly colored but much shorter billed and with spotted sides, feeds in flocks.

In Flight Entirely gray above and barred or white below, with plain wings and tail. All other rock shorebirds—turnstones, Surfbird, Rock Sandpiper—with conspicuous flight patterns. Nonflocking, flight often dashing and erratic and ascending.

Voice Flight call a fairly high-pitched, ringing series of notes often given and easily heard over sounds of surf. Only Whimbrel among common North American shorebirds has similarly loud, extended call, but its notes lower and farther apart and all on one pitch, whereas tattler's series gives impression of alternately slightly lower and higher notes. Sometimes gives series of calls in pairs—*too-ee, too-ee, too-ee*—which should be kept in mind when considering possibility of Gray-tailed Tattler. Breeding-ground song more complex with additional piping notes but all rhythmically repeated.

Behavior Feeds solitarily or loosely associated with rock shorebirds of its own or other species, picking invertebrates from substrate or probing into mats of mussels and algae. In quieter habitats, feeds in water. Roosting birds more likely to be found with other rock-frequenting species. Often teeters or bobs rear end like Spotted Sandpiper.

Habitat Gravelly streams from lowlands well up in mountains, also nearby tundra for breeding. Nonbreeders usually on rocky shores. Outside Americas, many on sand beaches and mudflats and even occasionally freshwater wetlands.

Range Breeds from northeastern Siberia to Alaska and Yukon. Winters on Pacific coast from California (rarely farther north) south to Ecuador and west all across Pacific to eastern Australia. Migrants confined to Pacific coast. Vagrant inland, east to Massachusetts.

38.1 Breeding Wandering Tattler. Rock-frequenting shorebird, all gray above with white supercilium, heavily barred below, and bright yellow legs. Long wings (well beyond tail tip) and primary projection. Nasal groove extends beyond halfway to bill tip. Playa del Rey, California, USA, Apr 2000 (Brian Small).

38.2 Nonbreeding Wandering Tattler. Like breeding but unbarred, breast and sides gray like upperparts. Ventura County, California, USA, Apr 1986 (Mike Danzenbaker).

38.3 Juvenile Wandering Tattler. Like nonbreeding adult but coverts with pale fringes, tertials often fringed and sparsely dotted, and breast with fine pattern. Playa del Rey, California, USA, Oct 1997 (Brian Small).

38.4 Breeding Wandering Tattler. Entirely neutral gray above. Nonflocking. Pescadero State Beach, California, USA, Aug 1996 (Mike Danzenbaker).

38.5 Breeding Gray-tailed (left) and Wandering tattlers. Differences in shade and intensity of barring readily apparent in direct comparison. Note barred uppertail coverts in Gray-tailed. Chiba Prefecture, Japan, May 1998 (Norio Kawano).

39 Gray-tailed Tattler

(Heteroscelus brevipes)

This species, an Asian vagrant in North America, is much like Wandering Tattler but can be distinguished at close range by plumage characters and always by voice. The pair represents a great example of closely related shorebirds that overlap in winter distribution but frequent different habitats.

Size Weight 108 g, length 28 cm (11 in), bill 3.8 cm, tarsus 3.2 cm.

Plumages Bill black, grading into gray or grayish brown or yellowish at base, legs yellow. *Breeding adult* with fairly fine gray barring on breast and sides. White supercilium and dark loral stripe distinct. *Nonbreeding adult* like breeding adult but lacking bars; breast gray, posterior underparts white. Loral stripe a bit paler in this plumage, contrasting less with supercilium. *Juvenile* similar to nonbreeding adult but with fairly conspicuous pale fringes on coverts, scapulars, tertials, uppertail coverts, and tail; typically with pale dots on edges of tertials and, in some individuals, tail edge. Breast slightly darker than that of adult, with indistinct fine light bars visible at very close range.

Identification Much like Wandering Tattler but slightly smaller and averages slightly paler, upperparts slightly browner; tail slightly paler than back. Primary tips extend just to tail tip in Gray-tailed, usually distinctly beyond in Wandering. In all plumages, Gray-tailed with more conspicuous white supercilium, although some variation in both species. In breeding plumage much less heavily barred below, barring absent from belly and undertail coverts. Uppertail coverts barred (unbarred in Wandering), visible at close range if wings lifted. In nonbreeding and juvenile plumages, sides whitish, narrowly gray especially toward front, rather than extensively gray (best distant field mark), breast slightly paler in Gray-tailed. In juvenile plumage same differences evident, and in Gray-tailed upperparts more heavily dotted and fringed with white, with prominent white notches on tertials which are usually lacking in Wandering. To test your optics, look for structural differences: nasal groove shorter in Gray-tailed (extending about half bill length, compared with about two-thirds in Wandering), and scales on tarsi scutellate in Gray-tailed and reticulate in Wandering. A tattler away from rocks in North America is worth checking out, but bear in mind that Wandering occasionally wanders onto mud and sand substrates for foraging.

In Flight Gray above and white (barred in breeding plumage) below with plain wings and tail, as in Wandering Tattler. Slightly paler rump and tail of Gray-tailed may be evident in flight, but flight calls in these noisy birds are best means of identification.

Voice Flight call a fluid, double whistle—*tuweet*—quite different from Wandering's series of notes and reminiscent of Semipalmated Plover. May be extended to three-noted, rolled *tuduweet, tuweet-weet*, or doubled *tuwee tuwee*. As in Wandering, commonly calls when flushed or flying past.

Behavior Much like Wandering Tattler; feeds much more commonly on mudflats but also inhabits rocks. Forages individually but seems to aggregate in larger flocks than Wandering where common, although flocking unlikely to be observed in North America.

Habitat Breeds in habitats much like that of Wandering Tattler, shores of streams and lakes in hilly tundra. Nonbreeders on exposed sandy or rocky coasts, much more likely to be on sand beach or mudflat than Wandering.

Range Breeds in eastern Siberia. Winters in Southeast Asia and on Southwest Pacific islands south to Australia. Migrants mostly coastal in between, regularly to western Alaska. Vagrant farther south on North American Pacific coast.

39.1 Breeding Gray-tailed Tattler. Slightly paler than Wandering Tattler, center of lower breast and belly unbarred. Wings project just beyond tail tip. Nasal groove extends less than halfway to bill tip. Shiokawa, Aichi Prefecture, Japan, May 2003 (Tadao Shimba).

39.2 Nonbreeding Gray-tailed Tattler. Like breeding but unbarred. Sides with less gray than Wandering, supercilium often more conspicuous. Nasal groove longer than is typical. Broome, Australia, Apr 2003 (Clive Minton).

39.3 Juvenile Gray-tailed Tattler. Like nonbreeding adult but scapulars, coverts, tertials, and even tail with fringes and dots, distinctly more patterned than Wandering. Again, note lack of gray on side. Broome, Australia, Nov 2000 (Clive Minton).

39.4 Breeding Gray-tailed Tattler. Like Wandering Tattler but slightly paler, tail slightly paler than back. Flight calls only sure distinction. At close range, barred uppertail coverts of this plumage visible. Sakai River, Aichi Prefecture, Japan, May 2003 (Tadao Shimba).

40 Common Sandpiper

(Actitis hypoleucos)

The Old World counterpart of Spotted Sandpiper, this species wanders to Alaska from Siberia and should be sought on the Atlantic coast as well. Unspotted in breeding plumage, it is much more similar in its other plumages. Unlike Spotted, this species is not known to be polyandrous.

Size Weight 48 g, length 18 cm (7 in), bill 2.5 cm, tarsus 2.4 cm.

Plumages Bill pinkish at base, grading to black tip; legs olive to dull yellow. *Breeding adult* with fine dark shaft streaks on many feathers and fine black bars across many scapulars, coverts, and tertials. Breast finely streaked with black. *Nonbreeding adult* with bill gray at base, legs dull yellow. Like breeding plumage but back plainer, without shaft streaks, coverts with fine pale bars, and breast plain brown without streaks. *Juvenile* with bill and legs as in nonbreeding adult. Coloration similar but coverts more heavily barred and tertials extensively dotted with buff, tail fringed with alternating black and buff dots. Some mantle feathers and scapulars also fringed with buff in more heavily marked individuals.

Identification Most like Spotted Sandpiper but differs in numerous small ways. Only apparent shape difference is longer tail, but this a good field mark. Difference in tail length in species only about 0.5 cm, but makes tail project considerably farther beyond wing tips in Common (distinctly greater than half bill length) than in Spotted (no more than half bill length). Apparent length may vary depending on how bird standing, so extended scrutiny advised. In *breeding* plumage, Common lacks spots on underparts and instead has dusky brown breast, finely streaked with black, and might be mistaken for Solitary Sandpiper if behavior and shape didn't give it away. Common also more streaked above and with dark markings more regularly spaced than in Spotted. Common has duller bare parts, bill generally dark but often with pinkish base, legs olive or dull yellow; both bill and legs bright yellow in breeding Spotted. In *nonbreeding* and *juvenile* plumages, differences in bare parts less, but Spotted still averages brighter than Common. Juveniles show numerous small field marks, and if entire suite can be seen, identification should be guaranteed. Juvenile Common has dark dots on outer edges of tertials, usually (but not always) lacking in Spotted. Wing coverts in Common less strongly barred than in Spotted, not greatly different from similar barring elsewhere on upperparts; in Spotted, black and buff barring on coverts distinctly more conspicuous than any pattern on back. Brown patches on sides of breast often streaked and/or barred, those on Spotted paler and unmarked. Nonbreeding adult Spotted colored like Common, so age must be determined to use this mark. Distinguished from other small sandpipers in same ways as Spotted.

In Flight Brown above and white below, with conspicuous white wing stripes and less conspicuous white rear wing edge and whitish tail tip and edges. In flight, species distinguished by wing pattern; white wing stripe extends to wing base in Common but falls short of it in Spotted. This (especially with photo) would be excellent way to confirm sighting of a Common in North America. Flight style like Spotted, shallow wing beats with wings fluttering below horizontal, but flies more like other shorebirds in migration.

Voice Flight call a series of *weet* or *pee* notes, more like that of Wood Sandpiper than Spotted. Multiple notes of Common loud and shrill, less musical and higher pitched than those of Spotted, with terminal consonant of each note less evident. Notes usually given in threes or fours, those of Spotted often in pairs, but total overlap. Each has additional calls not typical of other.

Behavior Much like Spotted, with incessant teetering of rear end and occasional bobbing of front end. Feeds by picking, much insect eating. Like Spotted, territorial in winter as well as breeding season. Although thought of as solitary, forms flocks in migration, and sometimes aggregates at roosts, unlike Spotted.

Habitat Breeds on wide variety of freshwater lakes, streams, and rivers in many habitat zones; prefers rocky and gravelly shorelines to sandy or muddy ones. Nonbreeders on both fresh and salt water in most shorebird habitats; exceptionally broad habitat tolerance.

Range Breeds across Eurasia from U.K. not quite to Chukotski Peninsula, with wide latitudinal range extending well south of most other sandpipers. Winters primarily in sub-Saharan Africa, India, Southeast Asia, Indonesia, and Australasia. Migrants anywhere in between, including through western Aleutians. Vagrant elsewhere in Alaska.

40.1 Breeding Common Sandpiper. Much like Spotted Sandpiper in shape and behavior but bill and legs darker, duller. Upperparts and breast with fine black streaks, tertials finely barred, but lacking underpart spots of Spotted. Tail slightly longer than in Spotted, difference not always obvious. Borgefjell, Norway, Jul 1988 (Mike Danzenbaker).

40.2 Nonbreeding Common Sandpiper. Like breeding but lacks sharp streaks on upperparts and breast; tertial bars still evident. Montpellier, France, Dec 2001 (Thomas Roger).

40.3 Juvenile Common Sandpiper. More heavily patterned above than nonbreeding adult. Like juvenile Spotted but entire upperparts fringed and dotted, much more extensive than in Spotted; heavily dotted tertials diagnostic. Salo, Finland, Jul 1993 (Henry Lehto).

40.4 Nonbreeding Common Sandpiper. Like Spotted, flies with fluttery flight, wings below horizontal, near water surface. White wing stripe more extensive than in Spotted, reaches wing base; also more white in outer tail feathers (not visible here). Fuerteventura, Canary Islands, Dec 2003 (Göran Ekström).

40.5 Breeding sandpiper wing specimens. Common (above) with longer white wing stripe. Common Sandpiper, Sverdlovskaya, Russia, Jun 1994; Spotted Sandpiper, Duckabush River, Washington, USA, Jun 1992.

41 Spotted Sandpiper

(Actitis macularius)

The Spotted is unique among common North American shorebirds in behavior, running along shores with constant teetering motions and flying low across water with fluttering wings. It is one of the most ubiquitous breeding species all across the northern part of the continent and similarly ubiquitous as a nonbreeder on salt water and along rivers to the south. Often polyandrous, females defend territories against other females and mate with more than one male, the males furnishing all parental care.

Size Male weight 41 g, female weight 48 g, length 17 cm (6.75 in), bill 2.4 cm, tarsus 2.4 cm.

Plumages Slight size dimorphism (female larger). *Breeding adult* with bill pinkish to yellow-orange to yellow, with clearly defined black tip; legs yellowish to orange. Variably and in some birds profusely marked with dark brown streaks and bars above, heavily spotted with black below. *Nonbreeding adult* with bill yellowish to pinkish to horn color, rarely black. Legs may be bright as in breeding adults or duller, greenish or gray. Upperparts plain brown, many feathers with dark streaks, coverts finely barred with black. Supercilium slightly less distinct than in breeding adult, so eyering more conspicuous. Underparts white with pale gray patches on either side of breast. Typically a few black spots retained on undertail coverts and more rarely elsewhere. *Juvenile* with bill and leg colors as in nonbreeding adult. Like nonbreeding adult but no streaks on upperparts or dots on underparts, wing coverts more strongly barred buff and black. Some scapulars and tertials tipped or fringed with pale buff, often with fine dark subterminal bar. Tertials on some individuals sparsely dotted with dark brown.

Identification Unmistakable in breeding plumage. More similar to other small sandpipers in nonbreeding plumage, but behavior singles it out. If not behaving, distinguished by entirely brown upperparts, brownish breast, and white underparts with white mark extending up in front of wings. Somewhat Dunlin-colored but with shorter, straight bill and pale legs; calidridine sandpipers have pale feather edgings lacking in Spotted. Could be mistaken for Solitary or much rarer Green Sandpiper because of uniformity of brown upperparts. Rare Common Sandpiper most similar species.

In Flight Brown above and white below, with conspicuous white wing stripes not reaching either end of wing (in most shorebirds, wing stripe reaches wing base). Tail with inconspicuous white edges and tip. Characteristic fluttery flight, shallow wing beat with wings held below horizontal and usually low over water. Birds in migration may fly like other shorebirds but not known.

Voice Flight calls loud, repeated whistles, high pitched and often disyllabic: *peet weet weet weet* or *tuweet tuweet weet weet* being only two variations among several. Often series drops in pitch and volume toward end. Given commonly in summer and much less so in migration. Among regularly occurring North American species, only

Solitary Sandpiper has similar call. Adults on breeding grounds sing extended *weet* series and *pit-a-weet, pit-a-weet,* etc., as part of courtship display.

Behavior As solitary as Solitary Sandpiper, individuals defending territories along coastlines and rivers throughout winter. Recognized by constant teetering of rear end, also bobs front end like other tringines. Captures most prey by picking, often a jab into vegetation with bill horizontal; insects prominent in diet.

Habitat Lakes, ponds, rivers, streams, reservoirs, and open marshes for breeding—just about any freshwater habitat from lowlands well up into mountains. Nonbreeders on all freshwater and coastal habitats, including salt marshes, mangrove swamps, beaches, and rocky shores.

Range Breeds all across North America from north edge of forest south to northern California, southern Rocky Mountains, central Plains, Appalachians, and Virginia coast. Winters on Pacific coast north to Washington, Atlantic coast north to South Carolina, and from southern parts of southernmost tier of states south through Middle America and West Indies to Chile and Argentina. Migrants anywhere.

41.1 Breeding Spotted Sandpiper. Small sandpiper that incessantly teeters rear end up and down. Unmistakable when heavily spotted below in this plumage. Bill and legs brightly colored. Rocky Mountain National Park, Colorado, USA, Jul 2002 (Brian Small).

41.2 Nonbreeding Spotted Sandpiper. Like breeding but with no spots, although some individuals retain a few, especially on lower belly and undertail coverts. Bill darker and duller. Coverts with black bars. Extension of white of underparts in front of wing characteristic mark of species in all plumages. Ventura, California, USA, Mar 1999 (Brian Small).

41.3 Juvenile Spotted Sandpiper. Like nonbreeding adult but coverts extensively barred with both buff and black; scapulars and tertials sparsely marked (unmarked in adult). Seattle, Washington, USA, Aug 2003 (Stuart MacKay).

41.4 Breeding Spotted Sandpiper. Flight fluttery, wings below horizontal, and usually near ground or water; migrants may fly more like other sandpipers. Wing stripe reaching neither base nor tip of wing is characteristic. Seattle, Washington, USA, Jul 1983 (Dennis Paulson).

42 Terek Sandpiper

(Xenus cinereus)

This unmistakable species, a vagrant from Eurasia to North America, is easily recognizable in any plumage by its almost frantic foraging behavior and combination of long, upcurved bill and short legs.

Size Weight 73 g, length 20 cm (8 in), bill 4.7 cm, tarsus 2.9 cm.

Plumages Bill black, sometimes with light brown base; legs orange-yellow to bright orange. *Breeding adult* with blackish streaks on many feathers. Upper scapulars vary from black spotted to entirely black, forming pair of wide black lines down either side. *Nonbreeding adult* differs by reduction or lack of black streaks on upperparts and black markings on scapulars. Uppertail coverts less heavily marked than in breeding plumage, breast less conspicuously streaked and usually paler, and head markings may be less distinct. *Juvenile* averages slightly darker than adult, and many feathers of upperparts, in particular coverts, narrowly fringed with gray-buff. Black scapular markings often present but less well developed than those of breeding adults. Distinguished from autumn adults as much by unworn tertials as by minor plumage differences.

Identification Long upturned bill unique for small sandpiper, as is plain gray plumage with scattered dark markings. Normal foraging gait also a good field mark, but of course some birds will be standing still. Nevertheless, no other bird anywhere near its size looks like it. Roosts with Gray-tailed Tattlers where common, similar enough that bill must be seen to distinguish them in nonbreeding plumage.

In Flight Gray-brown above and white below, with conspicuous white rear border of wing. Outer wing pattern often contrasty, with blackish primaries and gray lesser and median coverts. White stripe on rear edge of wings distinguishes Terek from tattlers and other sandpipers in flight (only much larger and long-legged Common Redshank has similar, even more conspicuous, pattern). Flight at times fluttery and low, suggesting Spotted and Common sandpipers, at times more rapid and erratic, with deep wing beats.

Voice Flight call a rolling series of short whistles, like a subdued Whimbrel but with fewer notes and quite different from multiple call of Wandering Tattler. Also an ascending, squeaky whistle, *pweeee-eeet*. Like other tringines, quite vocal.

Behavior Unusually active even for shorebird, running swiftly along shorelines and over flats with tail up and head lowered; looks as if it might tip forward at any second and plunge bill into mud. Frequent stops and starts and direction changes. Rapid gait allows it to capture large, exposed prey, mostly crabs, although also probes for same and feeds in water with avocetlike scything. Teeters like Common and Spotted, especially when disturbed. Roosts in large flocks with other species where common.

Habitat Breeds primarily in boreal forest, extends slightly into tundra and steppe on either side; prefers lush grasslands with shrubs and proximity of water. Nonbreeders mostly on coastal mudflats.

Range Breeds across Arctic from Finland to eastern Siberia. Winters on coasts of Africa, southern Asia, and Australasia. Migrants anywhere in between, including western Aleutians. Vagrant south on Pacific coast to Baja California, east to Manitoba, Massachusetts, and Barbados.

42.1 Breeding Terek Sandpiper. Tringine with long, upturned bill and short, bright yellow legs. Dark stripe on upper scapulars typical of this plumage. Often roosts with Gray-tailed Tattlers (behind). Shiokawa, Aichi Prefecture, Japan, May 2003 (Tadao Shimba).

42.2 Nonbreeding Terek Sandpiper. Like breeding but entirely plain above. Forages at a run with head low. Carmel, California, USA, Sep 1988 (Mike Danzenbaker).

42.3 Juvenile Terek Sandpiper. Like nonbreeding but fine fringes and scattered dots on upperparts; some have dark scapular stripe. Shiokawa, Aichi Prefecture, Japan, Sep 2003 (Tadao Shimba).

42.4 Juvenile Terek Sandpiper. Plain above, somewhat tattlerlike, except for distinctive white rear edge of inner wing, not as wide as in Common Redshank. Easily distinguished from that species by short legs and lack of white on back. Shiokawa, Aichi Prefecture, Japan, Sep 2003 (Tadao Shimba).

Curlews (Numeniini)

Long, decurved bills characterize this group of nine species (all of which have occured in North America), except for the aberrant Upland Sandpiper with a short, straight bill adapted for permanent life in the uplands. The curved bills may be a special adaptation to pull crabs and polychaete worms from their burrows. Another characteristic of the group is the lack of seasonal plumage change and minimal age differences in plumage. Females are slightly larger than males, with distinctly longer bills; juveniles have shorter bills which continue to grow through their first year of life. It would be interesting to know the reason for the strong sexual dimorphism in bill length.

43 Upland Sandpiper

(Bartramia longicauda)

This small curlew is the quintessential upland sandpiper, shunning water at all times and characteristic of waving-grass prairies and open fields. Its long neck allows it to survey the surroundings as it feeds, and its coloration is perfect to match the dead stalks of spring, autumn, and winter. Its song is the essence of the northern prairies.

Size Weight 150 g, length 27 cm (10.5 in), bill 2.9 cm, tarsus 4.9 cm.

Plumages Bill mostly yellowish, with tip and upper edge black; legs greenish yellow to fairly bright yellow. *Adult* scapulars and tertials with dark brown bars, former also with narrow buff fringes. *Juvenile* looks more striped than barred above because of conspicuous pale fringes. Most feathers of upperparts darker than those of adults, not showing same barred pattern but conspicuously fringed with pale buff or white. Tertials dotted with black just inside their pale fringes, whereas in adults crossbarring extends to edges of each feather. Otherwise age classes look much alike, although some juveniles washed with rich buff all over.

Identification Distinctive because of upland foraging; upright posture, long-necked and -legged look; brown, mottled plumage; and short, yellow-based bill. If bill were longer, could easily be mistaken for a curlew, and easily confused with vagrant Little Curlew if bill shape and size and details of coloration not closely seen. Other species that might occur with it—golden-plovers, Pectoral and Buff-breasted sandpipers—much smaller and differently colored.

In Flight Entirely brown above and largely white below, long tail with conspicuous dark and light barred edges. Curlewlike but smaller, and look for short bill; also, long tail gives distinctive shape. In comparison with some other brown shorebirds, rump entirely dark and strong contrast between dark outer wing and paler inner wing.

Complex tail coloration recalls snipe. On breeding grounds often flies with fluttery wing beats like Spotted Sandpiper, in migration with normal shorebird flight.

Voice Flight call Whimbrel-like but not so loud and with notes more run together, distinctly bubbling. Another flight call a liquid *putilip*, sort of a gurgled whistle. Breeding-ground song a long whistle, trilled at beginning, ascending and then descending, *whrrreeeee-wheeeeyuuuuuuu*, suggesting a wolf whistle.

Behavior Forages by moving steadily through grass and forbs and picking prey items or shallowly probing for them. Loosely social on breeding grounds, more so in migration, then territorial on winter grounds. Teeters rear part of body at times, not as actively as Spotted Sandpiper. Holds wings up on landing.

Habitat Medium to tallgrass prairie for breeding, at higher elevation in western part of range and in artificial grasslands such as airports in eastern part. Nonbreeders in similar habitats but typically shorter than for breeding, also in plowed fields and coastal dunes.

Range Breeds from Alaska southeast across western Canadian provinces to northern Great Plains, then across northeastern USA and far southern Canada to Atlantic coast. Western populations scattered and often isolated, declining. Winters in southern South America. Migrants anywhere east of Rocky Mountains south through eastern Mexico and Central America, much rarer farther west and rare through much of West Indies.

43.1 Adult Upland Sandpiper. Long-necked, small-headed, short-billed sandpiper of upland habitats. Brown at a distance, complex pattern at close range. Looks plain faced, big eyed; bill mostly yellow, legs yellow. Tail exceptionally long for sandpiper. Kidder County, North Dakota, USA, Jun 2002 (Brian Small).

43.2 Juvenile Upland Sandpiper. Like adult but scapulars and coverts darker centered, bordering fringe more conspicuous, so pattern is of pointed scallops. Aransas County, Texas, USA, Oct 1987 (Steven Holt/VIREO).

43.3 Adult Upland Sandpiper. Brown curlewlike sandpiper with short bill and long tail. Strong contrast between outer and inner wing. Texas, USA, May 1985 (L. Page Brown/ BioDiversity Institute).

44 Little Curlew

(Numenius minutus)

Fortunately, this species has not suffered the same fate as its relative the Eskimo Curlew, and it still strides around Australian grasslands and calls from spruce trees in Siberia. Shunning the mudflats and crab diet of other curlews, it is diminutive in size and has a short bill, which make it a good bridge to the un-curlewlike Upland Sandpiper.

Size Weight 150 g, length 27 cm (11 in), bill 4.4 cm, tarsus 4.9 cm.

Plumages Bill black, base of lower mandible reddish pink; legs pinkish to greenish to blue-gray, variation not obviously correlated with age. *Adult* tertials brown with pale edges and widely spaced narrow black bars. *Juvenile* very similar to adult but with conspicuous pale buff dots along edges of mostly black tertials. May be overlap; feather wear best criterion for separating juveniles from autumn adults.

Identification Very small curlew, much smaller than Whimbrel, and with short, not very conspicuously curved bill. Relatively long legged and long tailed for a curlew, actually as much like Upland Sandpiper as larger curlews, but coloration and curved bill easily distinguish it from that species. Upland Sandpiper a bit smaller, with still shorter, yellow-based bill, yellow legs, plain face with almost no hint of stripe through eye, and dark chevrons on sides. Basic pattern like Whimbrel, with conspicuous dark head stripes and similar bill color, but more washed with buff all over, often little or no barring on sides. Median crown stripe less conspicuous, dark loral stripe interrupted; legs more often pinkish. See Eskimo Curlew.

In Flight Plain brown above like small *hudsonicus* Whimbrel, with toe tips extending beyond tail tip; lacks buff bars on primaries characteristic of Whimbrel and Bristle-thighed Curlew. Shorter tailed and longer billed than Upland Sandpiper.

Voice Flight call reminiscent of Whimbrel but shorter, often with three notes, higher pitched and distinctly harsh or hoarse. Different from Upland Sandpiper's more musical, rippling call. Alarm calls sound like *coo-ee* or *cui-cui*, when repeated blend into typical flight call. See Eskimo Curlew.

Behavior Forages in upland habitats by walking briskly, capturing insects; stance usually quite erect except when capturing prey. In small to large flocks where common, singles in North America.

Habitat Breeds in openings in boreal forest. Nonbreeders in open fields and pastures.

Range Breeds in central Siberia. Winters in Southeast Asia and Australia. Migrants through eastern Asia. Vagrant to North America; one spring record from western Alaska, another from coastal Washington, and at least four from California in fall.

44.1 Adult Little Curlew. Small curlew patterned like Whimbrel but considerably smaller, much shorter billed. Central crown stripe less evident, dark loral stripe interrupted. Legs usually pinkish rather than gray. Bill curvature, more patterned head, and different bill and leg color sufficient to distinguish from somewhat similar Upland Sandpiper. Ishikawa Prefecture, Japan, Apr 1995 (Norio Kawano).

44.2 Adult Little Curlew. Pattern of scapulars and tertials more typical of juvenile plumage, yet appears to be replacing duller plumage. Curlews look the same throughout the year, yet show some unexplained variation. Carmel, California, USA, Sep 1994 (Brian Small).

44.3 Juvenile Little Curlew. White notches on tertials typical of this plumage, but distinction from adult may be guaranteed only by fresh plumage in fall, as in this individual. Darwin, Australia, Oct 1988 (Dennis Paulson).

44.4 Little Curlew. Small, plain brown curlew, with short, decurved bill. Shorter tail than Upland Sandpiper, less contrast between inner and outer wing. Like smaller edition of *hudsonicus* Whimbrel in flight. Carmel, California, USA, Sep 1994 (Mike Danzenbaker).

45 Eskimo Curlew

(*Numenius borealis*)

This small, stripe-headed curlew is the only North American shorebird that may be lost to us, its extirpation probably due to overhunting, habitat alteration, and the extinction of a grasshopper, one of its primary prey species. No shorebird sighting would be more exciting than a present-day encounter with and adequate documentation of this species.

Size Weight about 340 g, length 32 cm (12.5 in), bill 5.3 cm, tarsus 4.3 cm.

Plumages Bill black, limited reddish at base of lower mandible; legs gray, perhaps as variable as in Little Curlew. *Juvenile* much like adult but upperparts with broader pale (buff or cinnamon) edgings, underparts with more limited markings. Differences may not be striking in field, distinction from fall adult resting on unworn plumage.

Identification Rather like Little Curlew but also to be compared with *hudsonicus* subspecies of Whimbrel, which it resembles greatly. Smaller than Whimbrel, with shorter bill, but colored much like it. Juvenile Whimbrel has shorter bill than adult and would overlap with Eskimo; in fact, same-length bill on larger Whimbrel looks very small indeed. Differs from Little Curlew in slightly larger size and longer, more obviously curved bill; longer wings (extending beyond tail tip, not so in Little) and relatively shorter legs; less contrasty head pattern (crown stripe indistinct or lacking); and sides with chevron-shaped markings. In Eskimo Curlew, loral stripe reaches bill base, falls well short of it in Little Curlew, giving latter plainer-faced look.

In Flight Much like Little Curlew but with unbarred rather than barred primaries (often difficult to determine), darker (cinnamon vs. buff) underwings. Toe tips don't extend beyond tail tip. Strong buff coloration and unbarred primaries also distinguish from Whimbrel.

Voice Flight calls described as soft, repeated, melodious whistles, clearly not as loud and striking as those of Whimbrel. Not clear about distinction from Little Curlew but presumably less harsh.

Behavior An upland forager, striding about with upright posture much like Little Curlew. Like several other large shorebirds, uses berries as primary food to fatten for fall migration. Gathered in large flocks when abundant, as it was more than a century ago.

Habitat Breeding habitat treeless tundra, as far as known. Wintering and migration habitat open grasslands but also heath meadows, salt marsh, and even intertidal areas. Attracted to burned areas.

Range Bred formerly in northern Canada and probably Alaska. Wintered in southern South America. Migrants in fall to Maritimes and south to New York, then across Atlantic to South America, in spring north through midcontinent. Perhaps extinct; rare sightings continued to be reported to end of twentieth century, but last unequivocally documented records were photos taken in Texas in spring 1962 and a bird shot by hunter in Barbados in fall 1963.

45.1 Adult Eskimo Curlew. Probably extinct. Smaller and shorter-billed than Whimbrel but not by much (recall juvenile Whimbrels shorter billed than adults). Very like that species, differs by cinnamon underwings and unbarred primaries. Wings considerably longer than in Little Curlew, extend well beyond tail tip (also differs from Whimbrel in this way); sides more heavily marked and loral stripe complete. Galveston Island, Texas, USA, Apr 1962 (Don Bleitz/Western Foundation of Vertebrate Zoology).

46 Whimbrel

(*Numenius phaeopus*)

Plain brown at a distance, but there is nothing plain about the Whimbrel's personality. Alert and elusive, it nevertheless tames down on beaches where human encounters are frequent and is usually a conspicuous part of the flocks of large shorebirds that frequent North American coasts.

Size Male weight 355 g, female weight 404 g, length 41 cm (16 in), male bill 8.3 cm, female bill 9.1 cm, tarsus 5.8 cm.

Plumages Slight size dimorphism (female larger with longer bill). Bill entirely black (perhaps only on breeding grounds) or with variable amount of reddish at base of lower mandible, at maximum involving basal half. Variation in amount of pink at bill base needs to be better understood in all curlews. Legs dull blue-gray. *Adult* rather drably marked above; scapulars, tertials, and coverts gray-brown with darker brown bars and light brown notches. *Juvenile* with upperparts more brightly patterned than adult, with dark brown mantle feathers, scapulars, and tertials dotted and notched with pale buff to whitish. More contrast between scapulars and coverts in juveniles than in adults. Rump also more vividly barred (dark brown and buff) than in adult, contrasts more with tail.

Subspecies North American breeders and migrants entirely brown above with no contrasting white areas, easily distinguishable from other subspecies. Two subspecies distinguished within this complex, smaller *N. p. hudsonicus* from Hudson Bay and larger *N. p. rufiventris* from Alaska and northwest Canada. Many authors recommend splitting American as separate species (Hudsonian Whimbrel, *N. hudsonicus*) from Eurasian Whimbrel. Eurasian visitors to North America include *N. p. phaeopus* (breeds from Iceland to western Siberia, winters in Africa, vagrant all along North American Atlantic coast and inland to Ontario and Ohio, also West Indies), much whiter beneath and under wings and with conspicuous white rump and lower back and white-and-brown barred tail. *N. p. variegatus* (breeds in eastern Siberia, winters from India to Australia, vagrant to Alaska and down North American Pacific coast) like *phaeopus* but more heavily marked on underparts and under wings and more likely to be streaked on lower back and barred on rump and uppertail coverts (but some have virtually unmarked lower backs). Some *variegatus* heavily marked beneath, barred all across underparts. Variation in Eurasia clinal, so intermediate individuals likely.

Identification If bill not seen, distinguish from other large brown shorebirds by plain coloration, lacking reddish of Marbled Godwit and Long-billed Curlew, two common associates. If bill seen, only other curlews need be considered. Other common North American curlew, Long-billed, much larger and redder, with less distinct head stripes. Much rarer in most of North America, Bristle-thighed quite similar but differs in minor plumage characters, more distinctive in flight and by voice (see that species). Also see accidental Slender-billed Curlew, a much paler and more brightly marked species, and Little and Eskimo curlews, two other very unlikely small species that might be confused with Whimbrel.

In Flight American Whimbrels large and plain brown, white only on middle underparts. Plain brown color, with no reddish anywhere, and downcurved bill distinctive. Much smaller and shorter-billed than Long-billed and Far Eastern curlews, but see latter.

Voice Flight calls loud, somewhat hoarse, repeated whistles, *pi pi pi pi pi pi* (typically five to seven notes), distinct from all other North American species. Both Wandering Tattler and Upland Sandpiper have such repeated calls, but both have notes more run together and a bit more musical. Breeding-ground song loud, long rolling whistles and trills. Quite vocal in flight.

Behavior Forages equally well on sandy beaches and mudflats, walking methodically and pecking at mud or probing into burrows for crabs and worms; also often in rain-flooded fields.

Habitat Dry tundra for breeding. Nonbreeders mostly on coastal beaches and mudflats. Migrants often on grassland, plowed and flooded fields, even rocky shores. Regularly perches in mangrove trees at high tides.

Range Breeds in North American Arctic in Alaska and northwest Canada and around Hudson Bay, with apparent gap between these populations. Winters along southern coasts from California and Virginia south throughout Middle and South America but most common on north and west coasts of that continent; relatively rare in West Indies. Migrants anywhere, uncommon to rare in interior. See Subspecies for Eurasian range.

46.1 Adult Whimbrel (*hudsonicus*). Large, plain brown shorebird with decurved bill, standard for comparison of all other curlews. Strong head stripes distinguish from larger curlews. Adults with subdued pattern on upperparts, streaked neck and breast, and barred sides. American subspecies. Ventura, California, USA, Mar 1998 (Brian Small).

46.2 Adult Whimbrel (*hudsonicus*). Some individuals can be quite buffy. La Jolla, California, USA, probably Mar (Anthony Mercieca).

46.3 Juvenile Whimbrel (*hudsonicus*). Like adult except upperparts more strongly patterned, feathers darker with conspicuous white dots and notches. Bill typically shorter than in adult. Cape Spear, Newfoundland, Canada, Sep 1985 (Bruce Mactavish).

46.4 Adult Whimbrel (*phaeopus*). European subspecies much like American *hudsonicus* but underparts always pure white, not off-white as in *hudsonicus*, and thus bars on side very contrasty. Gambia, Jan 1988 (Richard Chandler).

46.5 Juvenile Whimbrel (*phaeopus*). Patterned upperparts distinctive from adult, as in American subspecies. Note relatively short bill. Cornwall, England, Sep 1998 (Richard Chandler).

46.6 Adult Whimbrel (*variegatus*). Siberian subspecies much like European *phaeopus* in having pattern slightly more conspicuous than American *hudsonicus*; some individuals heavily barred over entire underparts. Broome, Australia, Apr 1980 (Clive Minton).

46.7 Juvenile Whimbrel (*variegatus*). Patterning of scapulars and tertials, dark with pale markings, indicates juvenile. Heavily barred underparts typical of subspecies. Bulges perhaps from premigratory fattening. Mie Prefecture, Japan, Oct 2003 (Norio Kawano).

46.8 Adult Whimbrel (*hudsonicus*). Plain brown at any distance with conspicuous decurved bill; head stripes show up at closer range. Heavily barred underwings in this subspecies. Clearwater, Florida, USA, May 2002 (Wayne Richardson).

46.9 Adult Whimbrel (*phaeopus*). European birds with lower back pure white, underwings pale because dark bars relatively narrow. Similar to Eurasian Curlew but smaller, with shorter bill and striped head. Cleveland, England, May 2002 (Wayne Richardson).

46.10 Adult Whimbrel (*variegatus*). Siberian birds differ from *hudsonicus* in whitish lower back, somewhat less heavily barred underwings. Eastern subspecies of Eurasian Curlew with much whiter underwings and rump. Anou River, Mie Prefecture, Japan, May 2003 (Tadao Shimba).

47 Bristle-thighed Curlew

(*Numenius tahitiensis*)

This shorebird, with a tiny breeding range and far-flung winter distribution, is noteworthy in its surprisingly diverse foraging habits, its use of tools, and its flightlessness during molt. Its calls sound dramatically like a human whistling for immediate attention, and its warm colors suggest a Whimbrel bathed in the light of the setting sun, but the significance of its thigh bristles remains a mystery.

Size Male weight 374 g, female weight 428 g, length 41 cm (16 in), bill 9.5 cm, tarsus 6.0 cm.

Plumages Slight dimorphism in bill length and shape (female's longer, slightly thicker, less decurved). Bill black or with pink at base of lower mandible to entire base reddish up to half length of lower mandible and extreme base of upper (females typically with more pink than males, and both sexes probably more pink in nonbreeding season). Legs dull blue-gray. *Adult* with upperparts dark brown with bright to pale buff fringes and notches. Underparts bright buff, brown streaks on neck and breast and bars on sides. *Juvenile* may average more pale color on bill than adult, but more study needed. Similar to adult, but by October on wintering grounds much duller and darker above, with substantial percentage of light markings of upperparts worn off. Little indication of buff except on rump and tail. Remains on wintering grounds until at least three years old.

Identification Very like American subspecies (*hudsonicus* and *rufiventris*) of Whimbrel, a basically brown, midsized curlew with conspicuous dark head stripes. Bill varies surprisingly in thickness in both species, but legs in Bristle-thighed often look thicker than in Whimbrel. Differs from Whimbrel in overall brighter coloration, but only bright birds; dull birds much like Whimbrel. Even dullest Bristle-thighed somewhat buffy and patterned, rather like juvenile but very different from plain adult American Whimbrel. In bright Bristle-thighed, feathers of upperparts contrasty blackish and buff in comparison with Whimbrel's overall brown look. Bristle-thighed has sides less barred than Whimbrel, often no indication of bars, which are almost always visible in Whimbrel; however, some Whimbrels show reduced barring. Good look should reveal fine bristles hanging down from belly at base of legs on Bristle-thighed. Loud, characteristic vocalizations of both species will always furnish positive identification.

In Flight Much like Whimbrel—large and brown with curved bill and conspicuous head stripes—but easily distinguished by bright buff rump and tail, latter conspicuously barred. American Whimbrels unicolored above, and Eurasian subspecies of Whimbrel have white lower back and rump but brown tail. In Bristle-thighed, rump and tail look about same shade in flight.

Voice Flight call a three-noted whistle, *tee-oo-wit* or *whee-o-weet,* similar in cadence to familiar call of Black-bellied Plover but quite different in being crisp and sharp rather than smooth and plaintive. Sounds quite like human whistle. Call may be repeated several times by flying bird but is very different from Whimbrel's series of identical notes. Breeding-ground song derived from flight calls and includes many pure whistles, shifting up and down, for loud and dramatic display flights.

Behavior Flocks in migration, spreads out to feed. May forage like typical curlew, probing for invertebrates in soft substrates, but diet much more varied than in other species, as it feeds much in uplands, where it takes large crabs, scorpions, lizards, baby birds, and even mice and rats. Also steals eggs from breeding seabirds and uses stones to crack them open. Birds in both spring and fall in Alaska fill up on berries.

Habitat Hummocky upland tundra for breeding. Nonbreeders on sandy and rocky beaches and upland open areas on islands.

Range Breeds on Seward Peninsula and Yukon Delta in western Alaska. Winters on South Pacific islands from Hawaii south. Migrants across Pacific Ocean, only rarely seen on mainland except at Yukon Delta staging areas. Vagrant in spring along Pacific coast south to central California.

47.1 Bristle-thighed Curlew. Like Whimbrel but much more richly colored, buffy all over. Bold pattern of upperparts characteristic, dark feathers with wide buff fringes. Note bristles at base of legs, usually visible in good lighting. Juvenile and adult seem identical in field. Midway Atoll, Hawaii, USA, Mar 2001 (Brian Small).

47.2 Bristle-thighed Curlew. Best distinction from Whimbrel always bright buff rump, buff tail with narrow dark bars. Midway Atoll, Hawaii, USA, Mar 2001 (Brian Small).

47.3 Adult Bristle-thighed Curlew. Worn individual with upperparts much less patterned. Mie Prefecture, Japan, Jul 1978 (Teruaki Ishii).

47.4 Adult Bristle-thighed Curlew. Contrasty rump and tail furnish easy distinction from Whimbrel. Eurasian subspecies of Whimbrels have white up back. Primaries of Bristle-thighed not as strongly barred as in Whimbrel. Nome, Alaska, USA, Jun 1983 (Doug Wechsler/VIREO).

48 Far Eastern Curlew

(Numenius madagascariensis)

The largest of the sandpipers, this species is the Asian equivalent of the American Long-billed Curlew, with an amazingly long bill. White beneath instead of cinnamon, it is easily distinguished from Long-billed on its occasional visits to this continent.

Size Male weight about 600 g, length 58 cm (23 in), bill 15.0 cm, tarsus 8.0 cm; female weight about 800 g, length 64 cm (25 in), bill 22.0 cm, tarsus 9.0 cm.

Plumages Moderate dimorphism in size (female larger with distinctly longer bill). Bill black with varying amounts of pink to reddish on base of lower or both mandibles, at extreme on basal two-thirds; legs dull gray. *Adult* upperparts vary from gray-brown to warm reddish brown, feathers with darker central stripes and regular bars. *Juvenile* more distinctly marked above than adult. Scapulars dark brown with buff notches, tertials brown with buffy brown notches. Typically less striped below and with finer breast stripes than adult. Probably best distinguished in fall by plumage wear and typically shorter bill of juvenile.

Identification Most like Long-billed Curlew but differs from that species in lack of rich reddish coloration (but some reddish tinge on upperparts). Still room for confusion, as Far Eastern is buff to sandy brown, which might look reddish in some lighting conditions. However, Far Eastern much more obviously striped, both above and below, than Long-billed. Anything from neck and breast to entire underparts can be heavily striped in this species, whereas Long-billed has faint streaking on breast and upperparts more mottled than striped. See also rather similar Eurasian Curlew, not recorded on Pacific coast of North America but much more likely than Far Eastern on Atlantic coast.

In Flight Large with long bill, entirely brown. Shows none of reddish color that characterizes Long-billed, in fact looks more like Whimbrel, with heavily barred flight feathers, but considerably larger and longer-billed. Short-billed juvenile would have to be carefully distinguished from American Whimbrel in flight.

Voice Flight call a plaintive *curee, curee, curee* given fairly often, repetitiveness quite distinct from calls of other large curlews; also *cui-ui-ui-ui* alarm call, somewhat more musical than Whimbrel.

Behavior Where common, small flocks spread across mudflats probing for crabs and polychaete worms. Gather together in roosting flocks.

Habitat Wet meadows and grasslands for breeding. Beaches and mudflats during rest of year.

Range Breeds in eastern Asia. Winters in Southeast Asia and Australasia. Migrants mostly coastal, rarely to Bering Sea islands in spring and summer. Vagrant south along Pacific, one fall record in southern British Columbia.

48.1 Adult Far Eastern Curlew. Large brown curlew (the largest sandpiper) with very long bill. Pattern mostly of heavy striping above and below. This individual very worn, probably just arrived from Siberia. Broome, Australia, Aug 1992 (Clive Minton).

48.2 Adult Far Eastern Curlew. Some individuals of species are quite reddish. Streaks on underparts distinguish it from Long-billed Curlew, more reddish coloration and brown tail from Eurasian Curlew. Ibaraki Prefecture, Japan, Sep 2001 (Norio Kawano).

48.3 Juvenile Far Eastern Curlew. Age indicated by brightly marked fresh tertials with distinct notches, much shorter bill than adult. Shiokawa, Aichi Prefecture, Japan, Oct 1985 (Tadao Shimba).

48.4 Juvenile Far Eastern Curlew. Large, long-billed, plain brown curlew, lacking head stripes of Whimbrel, white rump of Eurasian, and cinnamon coloration of Long-billed. Relatively short bill indicates juvenile. Beidahe, China, Aug 2002 (Brent Stephenson).

48.5 Juvenile Far Eastern Curlew. Large, buffy brown curlew, distinguished from Long-billed by finely barred underwings not as bright reddish as in that species. Legs slightly longer than in Eurasian; more than half of toes project beyond tail tip. Beidahe, China, Aug 2002 (Brent Stephenson).

49 Slender-billed Curlew

(*Numenius tenuirostris*)

With breeding and wintering grounds never well known, this elegant curlew has become exceedingly rare. Although accidental in North America, it is very unlikely to occur again on this continent. It may be the next shorebird to follow the Eskimo Curlew to oblivion, a terrible shame in a time of heightened conservation awareness and management knowledge.

Size Weight about 360 g, length 39 cm (15 in), bill 8.2 cm, tarsus 6.5 cm.

Plumages Bill black, some birds with extreme base of lower mandible dark pink; legs gray to blue-gray. *Juvenile* has streaked rather than spotted sides but otherwise much like adult.

Identification Whimbrel-sized curlew, much paler and more strikingly marked than Whimbrel of American subspecies *hudsonicus,* with poorly defined stripes on head (darker crown contrasts with face), heavily black-spotted breast and sides, and very white underparts. Bill indeed slender, from base to tip, in comparison with other large curlews. Rather like *phaeopus* subspecies of Whimbrel but tail whiter, much whiter looking below, and lack of head stripes distinctive. More like Eurasian Curlew, but smaller and much shorter-billed than average Eurasian (but considerable size variation in all large shorebirds). In comparison with European subspecies (*arquata*) of Eurasian, sides spotted or streaked (Eurasian with chevrons and bars as well) and tail white with narrower dark bars (Eurasian usually with light brown on tail and more and wider bars). Eastern subspecies *orientalis* of Eurasian colored more like Slender-billed but even larger than western subspecies *arquata.* Smallest subspecies of Whimbrel, *N. p. alboaxillaris,* has similar breeding range to Slender-billed, could be mistaken for it; both extremely unlikely in North America.

In Flight Basically similar to but more contrasty and whiter than Eurasian Curlew and Eurasian subspecies of Whimbrel, very different from American curlews. Heavily spotted sides and relatively short, slender bill help distinguish from European relatives, also whiter tail base (fewer bars than in other species). Inner wing contrasts more with outer wing than in other species, both above and below, and secondaries with more conspicuous white markings.

Voice Flight call a *cur-lee* like larger curlews but sweeter and higher pitched than Eurasian Curlew, repeated more rapidly.

Behavior Forages with upright, even elegant, posture, walks and runs rapidly. Often crouches rather than flushing when raptors fly over.

Habitat Breeds in temperate marshes and bogs in boreal forest zone. Winters on coastal mudflats and salt marshes. Migrants often on meadows and pastureland near water.

Range Breeds (or bred) in southwestern Siberia. Winters (or wintered) in northern Africa, Mediterranean. Now so rare as to be in danger of extinction. Accidental in North America, one old record from Ontario.

49.1 Adult Slender-billed Curlew. Medium-sized curlew with indistinct head stripes, relatively shorter and very slender tipped bill, and heavily spotted underparts. Note very white tail. Merja Zerga, Morocco, Jan 1995 (Chris Gomersall).

49.2 Adult Slender-billed Curlew. Very white underwings, rump, and tail distinctive, as are conspicuous black spots on side and relatively short, slender, all-black bill. Merja Zerga, Morocco, Jan 1988 (Arnoud van den Berg).

50 Eurasian Curlew

(Numenius arquata)

A familiar sight on European moorlands, this large species wanders occasionally across the Atlantic and might make it across the Pacific some day. Its *cur-lee* call gives the group its common name.

Size Male weight 660 g, length 52 cm (20.5 in), bill 12.0 cm, tarsus 7.9 cm; female weight 790 g, length 55 cm (22 in), bill 14.5 cm, tarsus 8.4 cm.

Plumages Moderately dimorphic in bill length, female's longer. Bill black, variable amount of pink at base of lower mandible; legs blue-gray. *Nonbreeding adult* somewhat duller, grayer than breeding adult, plumage change more obvious than in Whimbrel or Long-billed Curlew. *Juvenile* more washed with buff, sides striped with no indication of spots or chevrons as in adult. Also shorter bill than adults of same sex.

Subspecies Atlantic coast subspecies, originating in Europe, is *N. a. arquata*. *N. a. orientalis* of Siberia, possible vagrant to Pacific coast, similar but somewhat paler overall, with significantly longer bill and slightly longer legs. In flight, underwings more extensively white than in European birds, whiter than those of Whimbrels from same region. Sides streaked in all plumages, not showing chevrons of adult *arquata*.

Identification Large curlew of Long-billed/Far Eastern size class. Much more like Far Eastern than Long-billed, but with whiter underparts, more heavily streaked sides, and no hint of reddish on upperparts.

In Flight Very different from Long-billed and Far Eastern curlews, with underwing coverts, lower back, rump, and tail mostly white. Even in silhouette, distinguishable from Far Eastern by shorter legs, less than half of toes projecting beyond tail tip (more than half in Far Eastern); Long-billed more like Far Eastern. Much like Eurasian subspecies of Whimbrel (note difference in subspecies) but underwing a bit whiter, less barred. Bill length and presence or absence of head stripes should be noted.

Voice Flight call a loud, somewhat melancholy, whistled *cur-lee*, often repeated, second note sharp and shrill; also a musical *cui-cui-cui-cui* away from breeding grounds. Quite vocal in flight.

Behavior Scatters across feeding area and strides around purposefully like other curlews, pecking and jabbing at and probing into substrate for worms, crabs, and snails. Forms large roosting flocks where common, flies in lines or vees.

Habitat Moors and marshes in boreal forest zone, also moist meadows in steppe and pastureland, for breeding. Coastal sand- and mudflats for nonbreeding, locally on muddy shores of large lakes and rivers.

Range Breeds across Eurasia from U.K. to Siberia, well short of Pacific coast. Winters from Europe and Japan south throughout Africa and southern Asia. Migrants throughout in between. Vagrant to North American Atlantic coast in spring, fall, and winter from Newfoundland to New York, also Bahamas; summer record from Nunavut.

50.1 Adult Eurasian Curlew (*orientalis*). Large, long-billed curlew with very white underparts, streaked sides. Whitish tail, cold gray-brown coloration, and often sparser streaking on underparts distinguish from quite similar Far Eastern. Considered adult because of primary molt. Asian subspecies, potential vagrant on Pacific coast. Isshiki, Aichi Prefecture, Japan, Sep 1996 (Tokio Sugiyama).

50.2 Adult Eurasian Curlew (*orientalis*). Very similar to Far Eastern Curlew, only good distinction in this photo whiter tail. Could be juvenile from contrast of tertials but bill seems too long. Okinawa Prefecture, Japan, Sep 1991 (Mike Danzenbaker).

50.3 Adult Eurasian Curlew (*arquata*). European subspecies, vagrant on Atlantic coast. Large, long-billed, heavily striped curlew. Presumably adult on breeding ground but does not show chevrons or spots on sides typical of many adults of this subspecies. Oulu, Finland, Jun 2003 (Henry Lehto).

50.4 Juvenile Eurasian Curlew (*arquata*). Like adult but more vividly patterned. Young bird with very short bill, reminiscent of Slender-billed Curlew but too buffy brown for that species. Loimaa, Finland, Aug 1990 (Henry Lehto).

50.5 Adult Eurasian Curlew (*arquata*). Large, long-billed curlew with white, mostly unbarred underwings. Underwings even less marked in *orientalis*, axillars white. Probably male by bill length. Durham, England, Jun 2003 (Wayne Richardson).

50.6 Adult Eurasian Curlew (*arquata*). Large, long-billed curlew with conspicuous white rump and lower back, tail paler than wings. Lack of head stripes distinguishes from European (*phaeopus*) Whimbrel. Probably male by bill length. Durham, England, Jun 2003 (Wayne Richardson).

51 Long-billed Curlew

(*Numenius americanus*)

With its long, curved bill and cinnamon coloration, this largest of our breeding shorebirds makes a striking symbol of western grasslands and coastal mudflats. It can be best appreciated when performing flight displays, at places its loud calls mingling with those of displaying godwits, willets, and snipes.

Size Male weight 640 g, length 51 cm (20 in), bill 12.7 cm, tarsus 8.1 cm; female weight 759 g, length 55 cm (22 in), bill 16.6 cm, tarsus 8.6 cm.

Plumages Moderate dimorphism in size (female larger with distinctly longer bill). Bill black, lower mandible reddish pink at base for about half length of bill; legs blue-gray. *Adult* with coverts cinnamon-buff with dark brown streaks and bars. *Juvenile* with tertials more brightly marked in fall than those of worn adults, with darker, wider stripes and bars on cinnamon-buff as opposed to grayish buff ground color. Coverts typically with wide teardrop-shaped marks along shafts and without bars or notches, quite distinct from adult markings on folded wings. Also on average finer breast stripes and more lightly striped breast than adult.

Identification Can be mistaken only for other large species. Whimbrel considerably smaller and shorter-billed (but overlaps with shortest-billed juvenile Long-billed) and has distinctive dark head stripes and no flavor of cinnamon. Marbled Godwit rather similarly colored but of course has slightly upcurved, pink-based godwit bill. If both sleeping at roost, curlew recognizable by larger size if comparison possible; also, curlew breast finely streaked, godwit unmarked or barred. See also Far Eastern and Eurasian curlews, both very rare in range of Long-billed.

In Flight Large with long, decurved bill, reddish brown body with quite reddish wings. Much richer colored than Whimbrel, most like Marbled Godwit because of reddish color of wings, but bill easily seen in flight.

Voice Flight call a loud *curlee*; not as vocal as Whimbrel under similar circumstances. During display flight on breeding grounds, loud, repeated *curlee* calls are followed by descending series of sharp whistles for song.

Behavior Birds stalk across mudflats and prairies singly but assemble in large roosts, often with other large shorebirds. Long bill used to extract crabs, shrimp, and worms from their burrows in mudflats, but equally adept at plucking insects (and nestling birds!) from fields.

Habitat Breeds in grasslands, mostly shorter grass, and extends to less-disturbed agricultural lands. Winters both on coastal mudflats and interior grasslands and pastures.

Range Breeds in Great Basin from southern British Columbia south to northern Nevada, across Rocky Mountains to western Great Plains from southern Alberta and Saskatchewan south to central New Mexico. Winters along Pacific coast from southern Washington south and on Texas and Florida coasts, also Central Valley of California and south through much of Mexico. Migrants throughout West and mostly on coasts in East. Rare in Central America, vagrant in West Indies and northern South America.

51.1 Adult male Long-billed Curlew. Large, long-billed, rich cinnamon-colored curlew. Distinguished from most similar species by unmarked reddish underparts, from Bristle-thighed by unstriped head and longer bill. Ventura, California, USA, Sep 1998 (Brian Small).

51.2 Adult female Long-billed Curlew. Female with distinctly longer bill than male, always distinguishable in adults. Ocosta, Washington, USA, May 2002 (Stuart MacKay).

51.3 Juvenile Long-billed Curlew. Distinguished from adult by striped wing coverts, with no crossbars. Bill shorter at first, grows rapidly to adult length in first winter. Ventura, California, USA, Oct 2001 (Brian Small).

51.4 Long-billed Curlew. Large and reddish, only curlew with virtually unmarked reddish underwings. Ventura, California, USA, Feb 2002 (Larry Sansone).

Godwits (Limosini)

Godwits are four species of large shorebirds, all of them found in North America, with long, slightly upturned bills and moderate to long legs. They often aggregate in large flocks while feeding with probing actions, more reminiscent of a clump of oblivious dowitchers than a spread of alert curlews. They often associate with curlews and willets in flight and regularly fly in lines. Three of the species change plumage dramatically, whereas the fourth, Marbled (the most common godwit in North America), looks more or less the same year-round.

52 Black-tailed Godwit

(Limosa limosa)

This is a familiar godwit of the Old World, common from Iceland to Australia, but the least common of the four species in North America. The long legs may be an adaptation for feeding in deepish freshwater, the rather straight bill for a foraging style much like that of a dowitcher.

Size Weight 291 g, length 36 cm (14 in), bill 7.8 cm, tarsus 6.6 cm.

Plumages Dimorphism slight in bill length (female's longer), moderate in plumage. *Breeding adult* with bill orange-pink with terminal third black; legs black. In brightest males, head, neck, upperparts, and breast rich reddish, belly and undertail white, barred with black. Complete supercilium. Female overall similar but less richly colored, rufous duller and with much white intermixed. *Nonbreeding adult* with legs dark gray. Overall plain, dark shaft streaks and faintly indicated pale fringes on coverts; no rufous or conspicuous barring. White supercilium extends back only to eye. *Juvenile* with legs dark gray; bill reddish at base, dark on top. Upperparts brown with buff to cinnamon fringes on all feathers, crown streaked and tertials notched with same color. Neck and breast cinnamon, rest of underparts white. Conspicuous white supercilium becomes buffy behind eye.

Subspecies Vagrants on Atlantic coast could be *L. l. islandica* from Iceland or *L. l. limosa* from western Europe, differing from one another in size (*islandica* with slightly shorter bill and tarsus) and color (breeding males of *islandica* more deeply colored with rufous extending farther onto belly; field differentiation possible). Vagrants on Pacific coast should be *L. l. melanuroides* from Asia, differing from western European birds by smaller size (bill averages about 2.5 cm shorter, tarsus about 1.5 cm shorter in *melanuroides*). It is more like *islandica* in breeding plumage and darker above than the other two in nonbreeding plumage.

Identification Rather like Hudsonian Godwit in all plumages because of vivid white wing stripes and black and white banded tail but with slightly longer legs and essentially straight bill, least recurved of any godwit. Breeding adults with unstreaked neck (diagnostic), likely to have reddish above (no trace of this color in Hudsonian), white lower belly and undertail coverts (reddish in Hudsonian), and more color at bill base. Juvenile much more richly colored than juvenile Hudsonian, with cinnamon neck; both show variably patterned upperparts. Nonbreeding adult much like comparable plumage of Hudsonian, but difference in wing length allows identification. In Hudsonian, wing tips project well beyond tail and primary projection obviously longer than bill-to-eye distance; in Black-tailed, wings project just beyond tail and primary projection no longer than bill-to-eye distance.

In Flight Large, with long, straight bill. Brown or rufous above, rufous below in breeding plumage, white below in other plumages. Vividly patterned, with striking white wing stripes and black, white-based tail. Underwings white. Differs from Hudsonian Godwit by white underwing coverts, broader white wing stripes that reach wing base (not to base in Hudsonian), and more white in tail base. All of toes and bit of tarsus project beyond tail in Black-tailed, less in other godwits. Willet and oystercatchers only other large shorebirds with vivid flight patterns.

Voice Not very vocal. Calls mostly single, low-pitched contact notes but also harsh, reedy double- and triple-noted flight calls, some of them the *kerreck* typical of godwits, when excited.

Behavior Where common, often in large flocks, but likely to be single in North America. Probes while wading like other godwits, commonly in fresh as well as salt water.

Habitat Breeds on wet meadows or hummocky moorlands, even reclaimed pastures and farmland, often near water. Nonbreeders on coastal mudflats and lagoons and freshwater marshes and flooded fields.

Range Breeds across Europe from Iceland and France to Russia, scattered from there east to eastern Siberia and south to Mongolia. Winters in southern Europe, northern Africa, southern Asia, and Australia. Migrants all across range, regularly to western Aleutians, more rarely farther east in Alaska. Vagrant to east coast of North America in migration and winter, records west to Ontario and Louisiana, south to Lesser Antilles.

52.1 Breeding male Black-tailed Godwit (*islandica*). Straight-billed, long-legged godwit with black and white tail. Iceland subspecies very dark and richly colored, rufous more extensive than in continental *limosa* and bill shorter. Bill much brighter than in Hudsonian. Selfoss, Iceland, Jul 1992 (Mark Peck).

52.2 Breeding Black-tailed Godwit (*limosa*). Continental subspecies duller and with rufous less extensive than in *islandica*. Sex of this individual not certain, dull male or bright female. Plain upperparts indicate female, prominent bars on side male. Montpellier, France, Mar 2003 (Thomas Roger).

52.3 Nonbreeding Black-tailed Godwit (*limosa*). Plain gray-brown above and on breast; bill pink rather than orange. Perhaps immature because no breeding color when others in flock already in that plumage. Montpellier, France, Mar 2003 (Thomas Roger).

52.4 Juvenile Black-tailed Godwit (*melanuroides*). Dark-spotted scapulars and notched tertials characteristic of this plumage, also overall buffy appearance. Distinguished from Hudsonian by longer wings, overall brightness, more conspicuous supercilium behind eye, pinker bill. Asian subspecies averages smaller than European ones. Aomori Prefecture, Japan, Sep 1998 (Mike Danzenbaker).

52.5 Juvenile Black-tailed Godwit (*melanuroides*). Vividly marked godwit with white wing stripe and black and white tail. Distinguished from Hudsonian by wing stripe reaching wing base and longer legs (tarsus visible beyond tail tip). Aomori Prefecture, Japan, Sep 1998 (Mike Danzenbaker).

52.6 Breeding male Black-tailed Godwit (*islandica*). From below, white underwings distinguish from Hudsonian Godwit, also straight bill and long legs. Sex and subspecies indicated by richly colored breast. Diksmuide, Belgium, May 2003 (Verbanck Koen).

53 Hudsonian Godwit

(Limosa haemastica)

Very localized in its breeding grounds, this species gathers in large flocks at a few staging areas in North America for its long flight to the southern end of the Americas. It is one of the more spectacular of the large shorebirds, with its rich summer colors and flashy flight pattern.

Size Male weight 225 g, bill 7.4 cm, tarsus 6.0 cm; female weight 289 g, bill 8.6 cm, tarsus 6.5 cm; length 37 cm (14.5 in).

Plumages Dimorphism slight in size (female larger with longer bill), moderate in plumage. Bill black with up to a bit more than basal half of lower or both mandibles pink; legs black. *Breeding adult* dark and mottled above, reddish and barred below. Female upperparts more mottled than in male, tertials often plain rather than brightly marked. Underparts paler chestnut, often intermixed with white but also barred as in male. *Nonbreeding adult* drab, light gray-brown on upperparts and breast; belly white. Dark loral stripe fairly conspicuous, white supercilium from eye to bill. Uniform at a distance; fine dark streaks on feathers of upperparts produce streaked effect only at close range. *Juvenile* bill with less pink than in adults, usually less than a third of bill. Browner than nonbreeding adult, with darker upperparts. Some mantle feathers and scapulars fringed, some scapulars and longer tertials prominently notched, and underparts washed with buff.

Identification Marbled Godwit larger and longer-billed, underparts cinnamon (Hudsonian underparts dark reddish, pale buff, or gray/white, depending on plumage). In nonbreeding and juvenile plumages, Hudsonian plain gray-brown or with faint scallops above, Marbled heavily marked and reddish above. Bar-tailed differs from Hudsonian in breeding plumage by lack of barring below, in nonbreeding and juvenile plumages by dark stripes and bars above. For a very similar species, see rare Black-tailed Godwit. Other than godwits, most likely mistaken for Willet, but latter has shorter, thicker bill with no hint of pink at base, supercilium not extending behind eye, pale tail, and gray legs.

In Flight Large, with long, straight bill. Brown above, rufous below in breeding plumage, white below in other plumages. Vividly patterned, with striking white wing stripes and black, white-based tail. Underwing coverts black. Very different from two other North American godwits because of broad white wing stripe, black underwings, and black-and-white tail, but much more like rare Black-tailed Godwit. Willet has even broader white wing stripe, so wing pattern more conspicuous than in godwit, but tail ringed black and white in godwit and all pale in Willet.

Voice Flight call *ta-wit,* like that of Marbled Godwit but higher pitched; not very vocal during migration. Song a long series of similar calls given during flight display on breeding grounds.

Behavior Probes deep in soft substrates for invertebrates, like other godwits and dowitchers. At times, aquatic-plant tubers make up large part of diet in this and some other shorebirds. Assembles in large flocks where common.

Habitat Breeds at taiga-tundra border. Winters on coastal beaches. Migrants on mudflats and interior lakes, reservoirs, and rice fields.

Range Breeds in disjunct populations from western and southern Alaska through arctic and subarctic Canada to, especially, south shore of Hudson Bay. Winters in far southern South America on both coasts, in only small numbers farther north to Peru and Uruguay. Migrants stage very locally in southern Saskatchewan and south shore of James Bay in fall, then make long flights across ocean to South America. Also common in eastern Great Plains in spring. Much less common elsewhere in North America; rare on Pacific coast south through Middle America and even rarer in Mexican interior and intermountain West. Rare in West Indies except Barbados in fall.

53.1 Breeding male Hudsonian Godwit. Godwit with black-mottled upperparts and dark chestnut, heavily barred underparts. Churchill, Manitoba, Canada, Jun 1989 (Brian Small).

53.2 Breeding female Hudsonian Godwit. Much duller than male with more gray above, white below. Harris County, Texas, USA, May 1997 (Brian Small).

53.3 Nonbreeding Hudsonian Godwit. Plain godwit with gray-brown upperparts and breast, whitish belly. This individual still in molt, much of breeding plumage replaced. Conneaut, Ohio, USA, Aug 2002 (Robert Royse).

53.4 Juvenile Hudsonian Godwit. Like nonbreeding adult but with faintly patterned scapulars, coverts, and tertials. Presquile Provincial Park, Ontario, Canada, Sep 1989 (Tadao Shimba).

53.5 Juvenile Hudsonian Godwit. Flight pattern indicated, with ringed tail and white wing stripe most prominent on outer wing. With Marbled Godwits. Tokeland, Washington, USA, Oct 2003 (Ruth Sullivan).

53.6 Juvenile Hudsonian Godwit. Only godwit, or for that matter large shorebird, with mostly black underwings. Breeding adults with reddish underparts. Willet has black underwings but much more conspicuous wing stripe, pale tail, and shorter bill and legs. Santa Clara County, California, USA, Sep 1990 (Mike Danzenbaker).

54 Bar-tailed Godwit

(Limosa lapponica)

This species, peripheral to most of North America, is nevertheless quite common in parts of Alaska, especially at its staging grounds on the Alaska Peninsula in fall. The birds feed until they bulge with fat, actually reduce the size of some of their digestive organs to conserve weight, then head for New Zealand, accomplishing the longest migration of any shorebird.

Size Male weight 282 g, length 40 cm (16 in), bill 8.2 cm, tarsus 5.4 cm; female weight 332 g, length 42 cm (16.5 in), bill 10.5 cm, tarsus 6.0 cm.

Plumages Dimorphism moderate in size (female larger with distinctly longer bill) and plumage. Bill mostly blackish with pink base; pink restricted to basal third and may be only on lower mandible. Legs black. *Breeding adult* typically brightly colored. Male dark brown and mottled above, unmarked rich chestnut below with some white on lower belly. Supercilium pale rufous, conspicuous eye stripe dark brown. Undertail coverts and often sides with black chevrons. Female much duller; supercilium whitish and upperparts much as in male but pale markings more extensive, in some producing striped look like that of nonbreeding adults and juveniles. Underparts vary from paler than that of male to almost entirely white, but more often with blackish bars on breast and chevrons on sides. *Nonbreeding adult* with pink at bill base more extensive than in breeding adult, extending almost halfway out bill; legs dark gray. Rather plain gray-brown above with darker brown feather centers producing overall striped look. Conspicuous white supercilium and dark eye stripe. Mostly white beneath, neck and breast gray-brown. *Juvenile* bill and legs as in nonbreeding adult but plumage more heavily marked. Upperparts brown, marked with buff to gray-brown notches, coverts predominantly buff with dark brown streaks. White supercilium conspicuous, dark eye stripe may be more conspicuous in front of or behind eye. Neck and breast finely streaked with brown and washed with buff; belly white, sides sparsely streaked and barred with brown.

Subspecies Vagrants to Atlantic coast should be European-breeding *L. l. lapponica.* Alaska-breeding birds and Pacific coast migrants are Asian subspecies *L. l. baueri,* differing from *lapponica* in slightly larger size and, especially, heavier markings; females more likely to be richly colored. In *lapponica,* lower back, rump, uppertail coverts, axillars, and underwing coverts mostly white; in *baueri,* these same areas heavily marked with brown. In field, *baueri* shows much less white in flight than *lapponica,* similar to difference between Siberian and European Whimbrels. One report of *baueri* from Massachusetts, weakening assumption that Atlantic vagrant shorebirds always come from Europe, Pacific ones from Asia.

Identification Most like Hudsonian Godwit in breeding plumage, differing in unbarred underparts and less pink at bill base. In both species sexual dimorphism must be taken into account. In nonbreeding plumage also like Hudsonian but with more patterned upperparts. Juvenile much like juvenile and nonbreeding Marbled but with more prominent pale supercilium and overall paler, but some Marbled, perhaps most worn individuals, can look quite pale. Bar-tailed whiter below, usually more contrast between slightly darker breast and paler belly, and faint streaks often visible on sides (not present in Marbled). Shape difference a good field mark, Bar-tailed with distinctly longer wings, primaries projecting well beyond tail (very little beyond it in Marbled), and primary projection longer (same length as distance from eye to bill base in Bar-tailed, distinctly shorter in Marbled). See rather different Black-tailed Godwit; these two species occur together commonly all across Old World but unlikely to be seen together in North America, as Black-tailed very rare visitor to normal range of Bar-tailed on this continent.

In Flight Only rather plain brown godwit with whitish lower back, showing neither rich reddish wings of Marbled nor black and white wings and tail of other two species. Legs slightly shorter than those of Marbled, so foot projection half of toe length (just about all of toe length in Marbled). Bill should be seen for elimination of brown curlews of same size such as Whimbrel (especially Eurasian subspecies) and Bristle-thighed.

Voice Flight calls include a low-pitched *kawee kawee* comparable to "godwit" call of Marbled. Not very vocal in migration. Breeding-ground song a repeated *awik awik awik* in flight display.

Behavior Probes rapidly in soft mud and water while advancing slowly, like other godwits and dowitchers. Large flocks during autumn staging in southern Alaska, a few at other times and places in North America.

Habitat Breeds on moist to dry arctic tundra, winters on coastal sand- and mudflats. Migrants primarily coastal, much less common in interior wetlands.

Range Breeds in western Alaska and locally across Siberia to Scandinavia. Winters widely across Old World from Africa to New Zealand. Migrants mostly along Eurasian coasts; very rare migrant, mostly in fall, south of Alaska on Pacific coast to Baja California. Vagrant to Saskatchewan and on north Atlantic coast of North America, also Virgin Islands and Venezuela.

54.1 Breeding male Bar-tailed Godwit (*baueri*). Only godwit with black and brown upperparts and entirely rufous, unbarred underparts. Minimal pink on bill in this species and plumage. Siberian subspecies is North American breeder. Nome, Alaska, USA, Jun 1998 (Brian Small).

54.2 Breeding female Bar-tailed Godwit (*baueri*). Unusually heavily barred individual; many females show some buffy to reddish below and less profuse barring. Broome, Australia, Mar 1995 (Clive Minton).

54.3 Breeding male Bar-tailed Godwit (*lapponica*). European subspecies much like Siberian and Alaskan *baueri*. Ruissalo, Finland, Aug 1998 (Henry Lehto).

54.4 Breeding Bar-tailed Godwit (*lapponica*). Females larger, distinctly longer billed, and much more plainly colored than males. Many look no different than nonbreeding plumage. Terneuzen, The Netherlands, May 2000 (Norman van Swelm).

54.5 Nonbreeding Bar-tailed Godwit (*lapponica*). Plain gray-brown godwit, upperparts more patterned than in Black-tailed and Hudsonian. Llanfairfechan, Wales, Mar 2002 (John Dempsey).

54.6 Nonbreeding Bar-tailed Godwit (*baueri*). Completing molt into this plumage, not all feathers replaced yet. Probably female from bill length. Mikumo, Mie Prefecture, Japan, Sep 2002 (Koichi Tada).

54.7 Juvenile Bar-tailed Godwit (*baueri*). Upperparts strongly patterned with white notches, breast buffy. Overall coloration much less reddish than in Marbled, also differs by more prominent supercilium, white showing on tail, longer wings and primary projection. Long bill indicates female. Tokeland, Washington, USA, Sep 2000 (Stuart MacKay).

54.8 Breeding adult Bar-tailed Godwit (*lapponica*). In flight, the only godwit with no wing stripe, conspicuously white rump and lower back. Varanger Fjord, Norway, Jul 1988 (Mike Danzenbaker).

54.9 Breeding male Bar-tailed Godwit (*baueri*). Lower back and rump heavily barred with brown in this subspecies. Buldir, Alaska, USA, May 2001 (Martin Renner).

54.10 Nonbreeding Bar-tailed Godwit (*baueri*). Underwing heavily barred in this subspecies. Molting into breeding plumage, probably male from bill length. Most New Zealand birds breed in Alaska. North Island, New Zealand, Feb 2003 (Brent Stephenson).

54.11 Juvenile Bar-tailed Godwit (*lapponica*). Underwing almost unmarked white in this subspecies. Yyteri, Finland, Jul 1987 (Henry Lehto).

55 Marbled Godwit

(*Limosa fedoa*)

This species looks more like a Long-billed Curlew than like its close relatives, but its bill shape and foraging behavior give it away as a godwit. It is the largest and most widespread and common of the godwits in North America. It can be the subject of close encounters on the Pacific-coast beaches it has long shared with human beachcombers.

Size Male weight 326 g, length 42 cm (16.5 in), bill 9.8 cm, tarsus 7.3 cm; female weight 391 g, length 44 cm (17 in), bill 12.0 cm, tarsus 7.8 cm.

Plumages Dimorphism moderate in size (female larger with distinctly longer bill), slight in plumage. Bill bright orange-pink on basal half, black at tip; legs black. *Breeding adult* with mantle feathers, scapulars, and tertials dark brown, notched or barred with buff. Coverts buff, striped and barred with dark brown. Breast and sides finely and variably (in some heavily) barred with brown. Males average more heavily barred than females, with brighter bill base. *Nonbreeding adult* with bill base pink, lacking orange tinge present in breeding season. Essentially like breeding adult but unbarred below or with sparse barring on sides. Breast usually with fine streaks. *Juvenile* like nonbreeding adult, no trace of bars on underparts. Coverts less heavily marked than those of adult, with poorly developed barring; typically mostly buff with brown central wedge or irregular linear markings. From a distance coverts look paler and plainer than back in juvenile, similar to back in adult. Breast usually unstreaked.

Subspecies Interior-breeding birds constitute subspecies *L. f. fedoa*, those of isolated Alaska population *L. f. beringiae*, known to migrate up and down Pacific coast south to southern California. Alaska birds slightly smaller than those from interior (wing length averages 2 cm shorter, bill and tarsus 1 cm shorter) but do not differ in plumage characters; can be distinguished only in hand.

Identification Large and reddish all over, standing out from contrastingly colored Hudsonian and much more likely to be mistaken for equally reddish Long-billed Curlew if bill can't be seen. About size of Whimbrel, with which it roosts regularly, but rich reddish color distinguishes it from that plain brown species. Feeding behavior entirely different (you'll never see curlews stitching mud as godwits do).

In Flight Reddish brown godwit with rich reddish wings and no conspicuous markings; size of Whimbrel and color of Long-billed Curlew but with almost straight bill. Flight feathers unmarked (those of Long-billed finely barred). See Bar-tailed Godwit.

Voice Flight call a moderately loud *kerreck kerreck*, enough like "godwit" to be mnemonically memorable but interspersed with chattering and laughing sounds in flocks. Breeding-ground songs a loud *ker-wick* and barking *radica radica radica*.

Behavior Typical godwit, associating in flocks and at times feeding close together, "stitching" mudflats by moving slowly forward and probing deeply and rapidly, much like dowitchers. May be plumage mimic of much more alert Long-billed Curlew, with which it often associates.

Habitat Breeds mostly on grassland, but peripheral populations on coastal tundra. Winters on mudflats. Migrates through similar habitats as well as freshwater lakes and reservoirs.

Range Breeds in northern Great Plains (Alberta to Manitoba and Montana to North Dakota) as well as small local populations in southwestern Alaska and along James Bay. Winters on coast from southern Washington and Middle Atlantic states south through Middle America, rarely to well south in South America; vagrant to West Indies. Migrants mostly coastal but can be temporarily common in some freshwater wetlands, especially in West.

55.1 Breeding Marbled Godwit. Largest godwit, rich reddish in all plumages. Heavily barred underparts characteristic of breeding. Benton Lake, Montana, USA, Jun 2001 (Brian Small).

55.2 Nonbreeding Marbled Godwit. Lacks barring on underparts, bill with more pink than in breeding. Long bill indicates female. Tokeland, Washington, USA, Sep 2000 (Stuart MacKay).

55.3 Juvenile Marbled Godwit. Very like non-breeding adult but coverts only faintly marked. Shorter bill could indicate male or not fully grown female. Tokeland, Washington, USA, Nov 2002 (Stuart MacKay).

55.4 Nonbreeding Marbled Godwit. Reddish all over, much like Long-billed Curlew except for bill. Castroville, California, USA, Aug 2003 (Glen Tepke).

Turnstones (Arenariini)

The two species of this small group are very alike, both small sandpipers with short, hard, pointed bills and fairly short legs for feeding and running on hard substrates. They are social, active, and noisy, among the most overt of the shorebirds and with exceptionally flashy flight patterns.

56 Ruddy Turnstone

(Arenaria interpres)

This is *the* turnstone, a shorebird so versatile that turning stones is only one of its many feeding strategies. Others include egg-eating in tern colonies and scavenging on carrion of all kinds. The turnstone bill and bright orange legs stand out in any flock, enhanced in breeding colors by the striking rufous and black pattern.

Size Weight 108 g, length 23 cm (9 in), bill 2.3 cm, tarsus 2.6 cm.

Plumages Moderate plumage dimorphism, sexes usually easily distinguished but variation in each sex. Bill black, legs bright to dull orange. *Breeding adult* with solid black breast markings, mostly rufous back. Many median coverts unusually large, look more like extra scapulars. Male crown white to light brown, heavily striped with black. Rufous and black markings on upperparts bold and contrasty, large swatches of pure colors. Female crown and hindneck brown, streaked with black; distinct from contrasty white-naped males. Upperpart pattern more irregular and mottled, rufous and black in smaller patches, giving overall brown tone. Single-sex flocks seen in migration. *Nonbreeding adult* with upperparts relatively plain and dark, many individuals showing scarcely a hint of ruddy. Head mostly brown, with faint indications of breeding-plumage pattern. Upperparts brown to blackish, most feathers fringed with gray-brown or dull rufous-brown. Breast markings mixed brown and black, bilobed at rear and enclosing pale area on each side of breast. Coverts dull rufous or gray-brown with darker central streak, medians smaller than in breeding. *Juvenile* typically duller but neater looking than nonbreeding-plumaged adult. Medium to dark brown above with scapulars, tertials, and coverts fringed with whitish and buff. Median coverts smaller than in adult, normal size. Pale fringes of juvenile gradually wear away, producing first-winter plumage darker and more uniform than in nonbreeding adult.

Subspecies North American subspecies *A. i. morinella* breeds from northeast Alaska across lower arctic Canada and migrates primarily down Atlantic coast but also less commonly in interior and along Pacific coast. Eurasian subspecies *A. i. interpres* breeds in eastern Canadian High Arctic and migrates to Europe. Northwestern Alaska breeders that migrate down Pacific coast are also *A. i. interpres* or intermediates between it and *morinella*. Amount of mixing away from breeding range poorly known. Typically *interpres* shows more black and less bright rufous above in breeding plumage than *morinella* and is overall darker in nonbreeding and juvenile plumages, with reduced pale markings. Both European and American Pacific birds average darker and less colorful than their counterparts on our Atlantic coast.

Identification Nothing looks like Ruddy Turnstone in bright breeding plumage, although its colors match those of Dunlin in same plumage. If bill apparent, birds in other plumages could only be confused with Black Turnstone, and darkest extremes of *interpres* subspecies (perhaps only in Eurasia) are colored surprisingly like that species but can be distinguished by bilobed breast patch, each lobe enclosing pale patch. Also, Ruddy always has white throat. Some Black Turnstones have brighter orange legs than others, but never as bright as those of typical Ruddy.

In Flight Chunky sandpiper with short bill and legs but unmistakable flight pattern: back, wings, and tail vividly marked with stripes and bars of white on brown or rufous. Patterned like Black Turnstone, Ruddy usually distinguished by overall brown and white rather than black and white appearance, but in poor lighting conditions, mistakes could be made.

Voice Flight call a series of notes rolled into low, semimusical chatter, unlike any other shorebird. Birds at times reduce chatter to few notes (*wee-ka wee-ka*) or even single note (*kir*), much higher pitched than chatter. Breeding-ground calls are similar chatters.

Behavior An active forager, living up to its name by tossing pebbles and seaweed aside with abandon, sometimes at a run, poking at other shorebirds that get in its way. Turns over surprisingly large objects to search for prey but also feeds on rock, sand, and mud substrates by direct picking. Sharp bill very efficient at disintegrating bird eggs and shellfish.

Habitat Fairly dry tundra (but usually near water) for breeding. Nonbreeding birds with minimal habitat specialization. Common on rocky, sandy, and muddy coasts, interior birds usually on lake shores.

Range Breeds across arctic latitudes from western Alaska to southern Baffin Island; also all across Eurasia. Winters on coasts from Oregon and New England south through Middle America and West Indies to Peru and northern Brazil, less commonly farther south to Tierra del Fuego; also in Europe, Africa, Asia, Australia, and Pacific islands. Migrants mostly coastal but regular in small numbers in interior of North America. Breeding birds from eastern Canadian Arctic Archipelago fly to Europe for winter; many from Alaska fly to Asia and South Pacific.

56.1 Breeding male Ruddy Turnstone (*morinella*). Bright rufous, black, and white sandpiper with short, pointed bill and orange legs. Median coverts very long in adult turnstones. North American subspecies. Manistique, Michigan, USA, May 2003 (Robert Royse).

56.2 Breeding male Ruddy Turnstone (*interpres*). Western birds average a bit more heavily marked with black than eastern ones, although subspecies differences in North America not well worked out, and Pacific birds may be intermediates between American *morinella* and Eurasian *interpres*. Note few nonreplaced median coverts. White feather on wing probably misplaced greater covert. Westport, Washington, USA, May 2003 (Stuart MacKay).

56.3 Breeding female Ruddy Turnstone (*interpres*). Females duller than males, with more black and brown and less bright rufous. Note dull plumage is in part because so many coverts and tertials are not replaced in spring. Nome, Alaska, USA, Jun 1998 (Brian Small).

56.4 Nonbreeding Ruddy Turnstone (*morinella*). Overall dull in this plumage, with very little rufous apparent. Again, note median coverts larger than in other shorebirds, droop over wing edge. Bill shape, leg color, and breast pattern easily distinguish it from all other shorebirds. Ocean County, New Jersey, USA, Jan 1996 (Mike Danzenbaker).

56.5 Juvenile Ruddy Turnstone (*morinella*). Neater plumage than nonbreeding, feathers more rounded and fringed with white or buff, median coverts not as large. Legs a bit duller. Distinction from other shorebirds as in nonbreeding. Conneaut, Ohio, USA, Sep 2002 (Robert Royse).

56.6 Juvenile Ruddy Turnstone (*interpres*). Eurasian juveniles average darker than American ones. American Pacific migrants may be intermediate. With Wandering Tattler, a common associate on Pacific islands. Okinawa Prefecture, Japan, Sep 2001 (Tomokazu Tanigawa).

56.7 Breeding male Ruddy Turnstone (*morinella*). Alternating rufous, black, and white produces one of flashiest shorebird flight patterns. Nonbreeding plumages show same pattern but brown instead of rufous. Unusual components include black patch between white back and tail base, as well as white patch outside scapulars (humeral feathers). Fort Myers, Florida, USA, May 2003 (Wayne Richardson).

57 Black Turnstone

(Arenaria melanocephala)

Another of the Bering Sea shorebirds, this turnstone has a minuscule range in comparison with its wide-ranging close relative. Its black and white color is nevertheless a common sight along the rocky Pacific coast of North America each winter.

Size Weight 122 g, length 23 cm (9 in), bill 2.4 cm, tarsus 2.7 cm.

Plumages Bill black, legs dull pinkish brown to orange-brown. *Breeding adult* with large white spot between bill and eye, hint of narrow white supercilium, and white flecks on sides of breast. Black on breast breaks up into large spots at rear on either side. Some median coverts unusually large, drooping from wings, and conspicuously edged with white. *Nonbreeding adult* browner than breeding adult, lacking white head and breast markings and thus uniformly colored on upperparts and breast. Some scapulars, tertials, and coverts narrowly edged with white, coverts large as in breeding plumage. *Juvenile* almost identical to nonbreeding adults but median coverts smaller, narrowly fringed all around with white, and tips of tail feathers slightly darker, not pure white. By midwinter, adults in fresh plumage with broad white tail tips, first-winter birds beginning to wear substantially and a bit paler than adults with white tail tips worn off (not easy to see in field).

Identification See Ruddy Turnstone for only other shorebird of same size and shape. Nonbreeding Black Turnstones could be mistaken for the three species that often accompany them on Pacific coast rocks: Surfbird, Wandering Tattler, and Rock Sandpiper. Other species paler gray in nonbreeding plumage, with yellow rather than orange-brown legs, and Surfbird and Rock Sandpiper have sparsely spotted sides. Surfbird larger with heavier bill, Wandering Tattler also larger, and Rock Sandpiper a bit smaller, last two with longer, slender bills.

In Flight Chunky, short billed, and short legged like Ruddy but with even more vivid pattern, same complex markings in black and white rather than brown and white. Other rock shorebirds not as vividly black and white as turnstone, with its white humeral and back stripes. Wandering Tattler plain gray, Rock Sandpiper gray with white wing stripes, Surfbird gray with white wing stripes and black-and-white banded tail.

Voice Flight call a shrill, high-pitched chatter, easily distinguished from Ruddy's lower, slower, and more melodic series of notes. Could be mistaken for Belted Kingfisher's rattling call. Breeding-ground calls include similar notes as well as series of rhythmically repeated "staccato" calls.

Behavior Not as active as Ruddy and not as likely to turn stones while foraging, as much of feeding is on solid rock substrates, where it picks among barnacles and mussels. However, birds in beds of shucked oysters or seaweeds prove just as proficient as Ruddy in this behavior, often pushing back algal mats. Assembles in good-sized flocks both for feeding and roosting.

Habitat Wet tundra or drier tundra near water for breeders. Rocky shores, including jetties, for nonbreeders, regularly spilling over onto sand- and mudflats.

Range Breeds in western Alaska. Winters on Pacific coast from southern Alaska south to Baja California, rarely to Nayarit. Migrants almost always along Pacific coast. Vagrant in North American interior to Montana and Wisconsin.

57.1 Breeding Black Turnstone. Black version of Ruddy Turnstone, with orange-brown legs; usually on rocks. This plumage features white markings on head and breast. Note long median coverts. Ventura, California, USA, May 1998 (Brian Small).

57.2 Nonbreeding Black Turnstone. Like breeding but very dark brown (looks black at a distance) rather than glossy black, no white markings on head and breast. Ventura, California, USA, Jan 1998 (Brian Small).

57.3 Juvenile Black Turnstone. Like nonbreeding but median coverts less enlarged. By midwinter, white tail tip wearing or worn off (intact on adults). Monterey County, California, USA, Nov 2002 (Robert Royse).

57.4 Black Turnstone. Flight pattern like Ruddy but black and white, with no brown. Blacker and more contrasty than Surfbird and Rock Sandpiper, common associates. Southern Oregon, USA, Jul 2001 (Jerry and Sherry Liguori).

Calidridine Sandpipers (Calidridini)

This group is the most diverse among North American sandpipers, representing all 24 species worldwide, as well as containing the smallest and most abundant species. These are the typical sandpipers that run along the beach and roost and fly in tight flocks and could be called "beachpipers" as a group. Feeding in a variety of environments nevertheless, they vary considerably in bill and leg length. Many are accomplished at both picking and probing.

58 Surfbird

(Aphriza virgata)

Surfbirds are known to few on their alpine-tundra breeding grounds, but they come down to spend the winter with Black Turnstones and other rock shorebirds all along the Pacific coast. They seem to be knots that evolved a ploverlike bill to forage on a solid substrate.

Size Weight 191 g, length 25 cm (10 in), bill 2.6 cm, tarsus 3.2 cm.

Plumages Bill black, lower mandible yellow at base; legs yellow. *Breeding adult* contrastingly patterned, with striped head and neck and heavily spotted sides and much of belly, spots chevron or heart shaped. Fresh scapulars vary from black with or without rufous notches to largely rufous with black subterminal blotches; most have white tips. Overall effect one of irregular rufous blotches, although some birds show no rufous at all. Bright rufous fades to pale buff or even white in returning adults in July; white tips to scapulars also worn off, these changes producing black-and-white-looking bird. *Nonbreeding adult* with upperparts and breast plain medium gray, throat and belly white. Only distinct markings white fringes on some coverts and lines of dark gray spots along sides, in some individuals extending across belly. *Juvenile* similar in overall coloration to nonbreeding adult but slightly paler. Head streaked and breast finely barred or scalloped, in contrast with more uniform coloration of adult. Scapulars with faint, narrow, dark subterminal bars and white tips; coverts more conspicuously scalloped with dark subterminal lines and white fringes.

Identification Most easily mistaken for its rock-inhabiting relatives in nonbreeding and juvenile plumages, this gray species distinguished by combination of short ploverlike bill, yellow legs, and spotted sides. In more brightly marked breeding plumage, looks like nothing else in North America except rare Great Knot in same plumage, and that species has gray legs in that plumage and much longer bill.

In Flight Short-billed, medium-sized sandpiper, gray backed and breasted (much more patterned in breeding plumage) and white bellied with conspicuous white wing stripe and pure white tail with black bar across tip. Distinguished from Black Turnstones with which it usually associates by larger size and plain back. Black and white tail distinguish it from all other medium-sized sandpipers; much smaller than similarly patterned Hudsonian Godwit.

Voice Typically rather quiet in nonbreeding range, vocalizations apparent only at close range. However, flocks sometimes very noisy with continuous soft, single, high-pitched notes and turnstonelike chatter (not as shrill as Black Turnstone). Complex and somewhat harsh song similar to those of other calidridines given in lengthy display flights.

Behavior Often forages with Black Turnstones and picks invertebrates (especially mussels and barnacles) like that species, but bill shape different and more likely to pluck prey from rocks rather than smash it as turnstone does. In much of North American winter range, usually smaller numbers in with Black Turnstones and rarely separate from them, but large roosting flocks occur northerly in migration route.

Habitat Alpine tundra for breeders. Rocky shores, including jetties, for nonbreeders.

Range Breeds in mountains of Alaska and Yukon. Winters on Pacific coast from southern Alaska south to Chile. Migrants only on Pacific coast. Vagrant east to Alberta and Florida.

58.1 Breeding Surfbird. Plover-billed, yellow-legged sandpiper of rocky habitats. Rufous-blotched scapulars and heavy spotting and scalloping on underparts characteristic of this plumage. Some spots heart shaped. Ventura, California, USA, Apr 1997 (Brian Small).

58.2 Breeding Surfbird. Distinctive pattern visible in front view. Note lack of webbing, typical of rock shorebirds. Pacific Grove, California, USA, Apr 1993 (Jim Rosso).

58.3 Breeding Surfbird. Rufous on scapulars faded to cream or even white by late summer. Ocean Shores, Washington, USA, Aug 1982 (Dennis Paulson).

58.4 Surfbird. Molting from breeding to non-breeding, difference in wear on feathers dramatic. Heavy wing molt with large gap in primaries indicating numerous primaries growing in. Princeton, California, USA, Aug 2000 (Peter La Tourrette).

58.5 Nonbreeding Surfbird. Plain gray plumage, sides spotted but not as heavily marked as in breeding. No other sandpiper has ploverlike bill, and no plover colored like this. Ocean Shores, Washington, USA, Oct 1981 (Dennis Paulson).

58.6 Juvenile Surfbird. Like nonbreeding adult but feathers of upperparts, especially coverts, with fine black subterminal line and white fringe; breast finely patterned. Ventura, California, USA, Sep 1998 (Brian Small).

58.7 Breeding Surfbird. Medium-sized sandpiper with conspicuous wing stripe, black and white tail. More contrasty than Rock Sandpiper, less so than turnstones. Only other flight pattern like this shown by much larger and longer-billed Hudsonian and Black-tailed godwits. Note faded markings on upperparts. Pescadero, California, USA, Jul 1996 (Mike Danzenbaker).

59 Great Knot

(Calidris tenuirostris)

Although looking like a large version of Red Knot, this large Siberian calidridine is perhaps the closest relative of the very different looking Surfbird. However, it forages instead on soft substrates with its probing bill and moves all the way to Australia in winter to do so.

Size Weight 155 g, length 27 cm (10.5 in), bill 4.3 cm, tarsus 3.5 cm.

Plumages Bill black, legs gray. *Breeding adult* looks heavily streaked, upperparts dark with whitish fringes and rufous markings on scapulars. Breast dark, solid or spotted black, sides marked with black spots and heart-shaped chevrons, undertail coverts sparsely black spotted. *Nonbreeding adult* with legs gray to greenish. Upperparts gray-brown, tertials darker brown. White supercilium may be almost complete or lacking either before or behind eye; usually does not reach bill base. Breast and sides irregularly streaked with brown, breast also with scattered larger, darker spots. *Juvenile* with legs as in nonbreeding adult. Upperparts brown striped, darker and more patterned than in nonbreeding adult. Supercilium usually complete. Breast suffused with buff and finely and evenly streaked or spotted with brown, sides similarly but more faintly marked.

Identification In all plumages, Great looks longer and sleeker than Red, more pointed toward rear; bill longer, slightly droopier. In bright breeding plumage, superficially like Surfbird but much longer bill, also longer neck and legs and very different foraging habitat. In nonbreeding plumage, much like Red Knot but larger, more distinctly streaked above, and spotted on breast. Face pattern different, with dark of lores extending upward to cut off supercilium from bill and supercilium and loral stripe overall less prominent. Typically, Red Knot has sparsely barred sides, Great sparsely spotted sides. Juvenile knot species quite different from one another, Great much more heavily marked on back and breast than Red; from distance, reminiscent of breast-belly contrast of Pectoral Sandpiper.

In Flight Moderately conspicuous white wing stripe and conspicuous white rump in all plumages. Longer winged than Red Knot, looking rangier in flight. Short legged, toes not projecting beyond tail tip. Between Black-bellied Plover and Red Knot in size, could be mistaken for either in flight, but shows contrasty white rump (because uppertail coverts white in this species, barred in others); in other two species entire tail looks paler than back. Both knots with more barring on uppertail coverts in breeding than in nonbreeding plumage.

Voice Flight calls low single or double whistles. Like its smaller relative, not very vocal, rarely calling when flushed.

Behavior Gathers in large flocks where common, much like Red Knots and sometimes with that species; perhaps a more active forager but very similar in feeding habits.

Habitat Breeders in alpine tundra, nonbreeders on mudflats and sandy beaches.

Range Breeds in Siberia. Winters in Southeast Asia and Australia. Migrants mostly along western Pacific shorelines. Vagrant to western Alaska, mostly in spring, a few records farther south to Oregon.

59.1 Breeding Great Knot. Medium-sized sandpiper (but large for calidridine) with medium-length bill. Rufous on scapulars and heavily spotted breast and sides characteristic of plumage. Heart-shaped spots unusual among shorebirds. Coloration like breeding Surfbird but bill, legs, and habitat very different. Broome, Australia, Mar 1995 (Clive Minton).

59.2 Nonbreeding Great Knot. Plain plumage, gray-brown above with finely streaked breast. Larger and longer-billed than Red Knot, sides spotted and streaked rather than barred. Probably first-year bird because no sign of breeding plumage on this date. Broome, Australia, Apr 1995 (Clive Minton).

59.3 Juvenile Great Knot. Much browner than nonbreeding adult, with heavily patterned upperparts and streaked breast. Note interrupted loral stripe and greenish legs. Much more heavily marked than juvenile Red Knot, could be mistaken for Pectoral Sandpiper. Chiba Prefecture, Japan, Sep 1998 (Mike Danzenbaker).

59.4 Juvenile Great Knot. Flight pattern white wing stripe, white rump, and darker gray tail. Mie Prefecture, Japan, Aug 2002 (Koichi Tada).

59.5 Nonbreeding Great Knot. Flight pattern distinctive, with narrow wing stripe and contrasty white rump. With Black-tailed Godwits, a common associate. Cairns, Australia, Dec 1987 (Göran Ekström).

59.6 Nonbreeding Great and Red knots. Larger size, longer bill, and rangier look of Greats evident in comparison with Red. Roosting with Black-tailed Godwits. Cairns, Australia, Nov 1998 (Dan Logen).

60 Red Knot

(Calidris canutus)

This is a large, short-billed calidridine, in breeding plumage variably patterned above and bright rufous below, red breasted enough to have been called "robin snipe." More than most migrant shorebirds, it aggregates in huge numbers at favored staging areas and is uncommon elsewhere. Dramatic declines in migrant knot populations in Delaware Bay may be attributable to overfishing of horseshoe crabs there.

Size Weight 135 g, length 24 cm (9.5 in), bill 3.6 cm, tarsus 3.2 cm.

Plumages Slight plumage dimorphism. Bill black, legs dark olive to dark gray or almost black. *Breeding adult* upperparts range from gray-brown (old) to variably marked; many are mostly black or black with large, paired subterminal rufous spots and white tips. Underparts bright rufous, white on belly varying from extensive to almost absent. Posterior sides and undertail coverts sparsely barred with black. Males average more brightly marked than females, with more rufous above; females also with more white below and more likely to retain barred winter feathers on breast. These are average differences, perhaps with total overlap. Fall adults can have largely blackish crowns and mantles as their pale fringes wear off. *Nonbreeding adult* with legs greenish, paler than in breeding adult. Upperparts and breast light gray-brown with faintly darker streaks, underparts with fine brown wavy bars on breast and sides. White supercilium and gray-brown loral stripe moderately conspicuous, ear coverts also dusky. *Juvenile* with legs greenish, may be paler than in adults. Coloration much as in nonbreeding adult but breast more likely to be less conspicuously marked and with fine brown streaks and dots instead of bars. Scapulars, coverts, and tertials with narrow, dark subterminal lines and white or buff terminal fringes. In younger juveniles, entire underparts may be washed with buff, persisting even into September in some individuals.

Subspecies *C. c. islandica* breeds in northeast Canadian Arctic and Greenland, winters in western Europe; in breeding plumage extensive rufous above, almost entirely reddish underparts. *C. c. rufa* breeds in lower-latitude Canadian Arctic, winters in southern South America (and southeastern USA?); upperparts with very little rufous, mostly gray contrasting strongly with underparts; paler below than *islandica*, with much white on lower belly and undertail coverts. *C. c. roselaari* breeds on Wrangel Island (wintering grounds unknown), and birds with similar coloration but intermediate in size between *rufa* and *roselaari* breed in western Alaska, thought to winter in Florida, Central America, and northern South America; like *rufa* but darker, again with more rufous above. Wintering grounds and migration routes of American subspecies still not adequately known. I cannot distinguish breeding-plumaged Red Knot specimens from Washington from those from Atlantic coast, so still some question of whether two coasts visited by different subspecies.

Identification Knots about dowitcher size and may be commonly mistaken for them, but much shorter bill always distinctive when seen. In sleeping flocks of breeding-plumaged birds, knots distinguished from dowitchers by back coloration, usually grayer (dowitchers brown), brighter rufous underparts, and extensively white lower belly (dowitchers entirely reddish or with white limited to undertail coverts, rarely with more extensive white). Knots in nonbreeding and juvenile plumages paler and grayer above than dowitchers, with whiter breast. More of a stretch to confuse them with any other common North American shorebirds, as Sanderling much smaller and Black-bellied Plover much larger, but both of these are associates. In breeding plumage, Curlew Sandpiper might be confused but is considerably smaller with fine, slightly curved bill. See Great Knot, rare in North America, for most similar species.

In Flight Medium-sized sandpiper with moderate-length bill. Heavily marked reddish, brown, and black above and rufous below (with white undertail) or entirely gray-brown above and white below. Inconspicuous white wing stripe, tail somewhat paler than back. Often with Back-bellied Plover, most like it in flight but less strikingly marked. Similar-sized dowitchers have white point up back and long bills sticking out front.

Voice Flight call a soft, melodious *cur-ret*, rather godwitlike. Most likely to be given by single birds in flight; not very vocal. Breeding-ground flight song a plaintive, repeated double whistle, *pooo-lee pooo-lee pooo-lee*, coupled with series of short trills.

Behavior Like dowitchers, knots often occur as dense feeding flocks, and the species often mingle, knots usually moving steadily and dowitchers often virtually stationary as they feed. Feeding by both picking and probing, also "plowing" into wet sand, especially for bivalves. Roosting and flying flocks, especially in Alaska and on Atlantic coast, can be very large.

Habitat Breeds on elevated arctic tundra. Migrates and winters on coastal sand beaches and mudflats; interior migrants usually on lakes and reservoirs.

Range Breeds locally in northwestern Alaska, Canadian Arctic Archipelago, and mainland north of Hudson Bay. Winters on Atlantic coast from Massachusetts (more commonly from Carolinas) south very locally to southern Argentina, rarely on Pacific coast of USA (north to central California) and south to Chile, much less common than on Atlantic coast. Migrants largely coastal, staging in spring in huge numbers at a few localities on each coast, but regular in small numbers across continent; much more dispersed in fall. Generally rare in West Indies, most regular on Barbados in fall. Also widespread in Old World, breeding in Greenland and northern Siberia and wintering on coasts of Europe, Africa, Southeast Asia, Australia, and New Zealand. See details under Subspecies.

60.1 Breeding Red Knot (*roselaari*). Rufous-breasted, medium-sized sandpiper with bill much shorter than in dowitchers. Pacific coast birds thought to be this rufous-backed, dark-breasted subspecies, but variation great all across continent. Ocosta, Washington, USA, May 2003 (Stuart MacKay).

60.2 Breeding Red Knot (*rufa*). Many breeding individuals with black streaks dominating upperparts, little rufous. In breeding range of *rufa* but not typical of subspecies. Cambridge Bay, Northwest Territories, Canada, Jul 2000 (Jim Richards).

60.3 Breeding Red Knot (*rufa*). Extensive white undertail characteristic of this species and most individuals in this plumage in North America. Many individuals of *rufa* with quite gray upperparts, pale rufous breast in full breeding plumage. Sanibel, Florida, USA, Mar 1996 (Brian Small).

60.4 Breeding Red Knot (*rufa*?). Color and pattern in this flock encompass variation that supposedly characterizes subspecies. Short-billed Dowitcher (behind knot on right) in same plumage for comparison. Knot less reddish brown overall, with conspicuous white posterior underparts; note also difference in leg color in breeding plumage. South Padre Island, Texas, USA, Apr 2002 (Anthony Mercieca).

60.5 Nonbreeding Red Knot (*rufa*). Plain light gray-brown in this plumage, with breast and sides lightly barred. Relatively short, straight bill important field mark. Note barred uppertail coverts. Sanibel, Florida, USA, Mar 1999 (Brian Small).

60.6 Juvenile Red Knot (*rufa*). Like nonbreeding but with fine black subterminal lines and white fringes around most feathers of upperparts. Conneaut, Ohio, USA, Aug 2003 (Robert Royse).

60.7 Juvenile Red Knot (*roselaari*). No obvious differences between Atlantic and Pacific coast juveniles, although thought to represent different populations. Difference in bill length may represent sexual difference (male shorter). Ventura, California, USA, Sep 1997 (Brian Small).

60.8 Nonbreeding Red Knot (*canutus*). White wing stripe in flight, rump and tail slightly paler than back; rather uniform looking at a distance, with less contrast between back and tail or rump than in Black-bellied Plover and Great Knot, somewhat similar species. Cleveland, England, Mar 2001 (Wayne Richardson).

61 Sanderling

(Calidris alba)

This is the sandpiper most people can picture, following waves out and preceding them in as it forages on high-energy beaches with its own high energy. The lack of a hind toe is indicative of its prowess as a runner. It is pale overall in winter but striking orange-rufous in breeding plumage.

Size Weight 52 g, length 19 cm (7.5 in), bill 2.5 cm, tarsus 2.6 cm.

Plumages Bill and legs black. *Breeding adult* orange-rufous on head, breast, and upperparts, although very variable, some individuals with little of that color, instead heavily marked with brown and black. Molts into this plumage late in spring and out of it quickly in fall. *Nonbreeding adult* above very pale gray-brown, with fairly distinct white supercilium; below entirely white. Uniformly pale appearance broken by dark brown eye, black bill and legs, and black lesser wing coverts, the last conspicuous when birds display at each other. *Juvenile* much like nonbreeding adult but heavily spangled or checkered with black above. Tertials darker than in adults, darkening to dark brown at tip and conspicuously fringed and, in some individuals, notched with white. White supercilium contrasts more with crown and eye stripe than in adult. Breast streaked on sides and may be washed with pale buff. Molting adults with black summer feathers scattered on upperparts (visible into September) can be confusing; however, pattern never regular like that of juveniles.

Identification In nonbreeding plumage, palest by far of small sandpipers at rest or in flight, only equaled in paleness by Snowy and Piping plovers and nonbreeding Red Phalarope. Sandpiper bill distinguishes it from plovers, lack of dark ear coverts from phalarope. In breeding plumage, orange-brown coloration distinctive, but shade similar to colors borne by Red-necked and Little stints, both smaller and with rich orange more restricted. Breeding-plumaged birds liable to be mistaken for juvenile Baird's Sandpipers because of similar size, black legs, and rather rich coloration of both species, but shouldn't overlap in time. Juveniles look somewhat black and white checkered above, different from any other small sandpiper, although some juvenile Spoon-billed Sandpipers similarly colored.

In Flight Pale gray-brown above and white below in nonbreeding plumage, slightly darker above in juvenile plumage, and reddish orange above and on breast in breeding plumage. White wing stripes conspicuous, enhanced by blackish lesser and median coverts. Darker rump stripe conspicuous as in other *Calidris*, rest of tail whitish. Has more conspicuous wing stripe than Dunlin, with which it often flies. Red Phalarope in flight somewhat similar but greater and median wing coverts not much darker than lessers, unlike Sanderling.

Voice Series of single, sharp *wik* or *kwik* notes given in flight or during aggressive encounters. Breeding-ground song rapid trills or chatters, some of them low pitched and sounding all too much like alarm clock.

Behavior Like little wind-up toys, run up and down beach following outgoing waves and outrunning incoming ones. Pick at surface of wet sand and probe deeply into dry sand, leaving beach covered with characteristic probe holes. Roost and fly in large flocks, but juveniles often forage away from dominant adults in autumn.

Habitat Generally dry arctic tundra for breeding. Sandy beaches for nonbreeding but will move onto tidal sand- and mudflats; shores of large lakes and reservoirs in interior.

Range Breeds on islands of central Canadian Arctic, less commonly on adjacent mainland and rarely in Alaska; also scattered across Eurasian Arctic. Winters on coasts from southern British Columbia and Massachusetts south through Middle America and West Indies to southern South America, much more common on Pacific coast of both Americas; also worldwide on temperate and tropical shores. Migrants mostly coastal but scattered across continent, with local spring concentrations; generally late in spring because of high-latitude breeding.

61.1 Nonbreeding Sanderling. Small sandpiper that runs and runs, usually into and out of waves. Short bill, black legs, very pale through most of year. In molt or at dull extreme of breeding plumage. Note lack of hind toe, adaptation for running (all other sandpipers have hind toe). Fort Myers, Florida, USA, Apr 2000 (Wayne Richardson).

61.2 Breeding Sanderling. Much orange-rufous on head, breast, and upperparts in full plumage; no other sandpiper colored exactly like this. Upper Coast, Texas, USA, May 2002 (Brian Small).

61.3 Breeding Sanderling. Bright extreme of this plumage; here probing in dry beach sand, where they leave characteristic marks. Ocean Shores, Washington, USA, May 2003 (Stuart MacKay).

61.4 Nonbreeding Sanderling. Pale in this plumage, with entirely white underparts. Black lesser coverts often visible, especially during aggressive displays. Orange County, California, USA, Jan 1995 (Larry Sansone).

61.5 Juvenile Sanderling. Like nonbreeding but upperparts, including crown, spangled with black. Breast pale buff at first, quickly fades to white. Dark tertials retained through first winter, allow age distinction. Ventura, California, USA, Sep 1997 (Brian Small).

61.6 Breeding Sanderling. Broad white wing stripe conspicuous in flight, more so than in other small *Calidris* with identical flight pattern. Orangey coloration also distinctive in this plumage. Clearwater, Florida, USA, May 2003 (Wayne Richardson).

61.7 Nonbreeding Sanderling. Palest sandpiper in flight, with broad white wing stripe; often in large flocks. Contrasting black lesser coverts show clearly in many birds. Few Dunlins in upper part of photo distinguished by longer bill, browner coloration, less conspicuous wing stripe. Ocean Shores, Washington, USA, Apr 1984 (Dennis Paulson).

Stints

The seven species of stints, all found in North America, are the tiniest of the sandpipers and have long been called "peeps" on this continent. Three things must be understood about their identification: (1) it can be very difficult, even with species occurring together regularly within a continent; (2) it is made even more difficult because all species wander with some regularity to the "wrong" continent; and (3) with care, it should be possible.

At first glance, all stints look cut from the same mold, but in fact there is quite a bit of variation in proportions and even size. Western is the largest, Least the smallest, and when these two are together there is little doubt about the size difference. The other species, however, form a series between the two extremes, and size is unlikely to be of any help distinguishing most of them. Bill length and shape are important characters, and the very long-billed Westerns, especially females, stand out from the pack, as do many Semipalmateds because of their relatively thick, blunt bill. Having said that, I caution observers that only female Westerns are definitively identified by their long, slightly droopy bill. All bills vary in length and shape to the point that the bill is only one of the numerous characters that must be noted for definitive identification of a stint. If Spoon-billed Sandpiper were considered a stint (it surely should be from most characteristics), its bill would suffice as a field mark.

Shape differences in stints are slight and are caused by relative wing length. Red-necked and Little have relatively long wings, with longer wing projections and primary projections, which makes them look quite pointed behind, but Western and Semipalmated are close behind them, with overlap. Least and Long-toed have virtually no primary projection, the wings and tail about equal, so they present a slightly stubbier look. Temminck's again looks rather pointed behind, but this is because of its long tail, often projecting beyond the wingtips.

It is important to note leg color to distinguish black-legged from pale-legged species, but black-legged ones have olive legs when very young, and there are records of "yellow-legged" species having dark legs. Thus this character, like so many others in stint identification, should be used with a grain of salt (or sand, if at the beach). While checking out leg color, note whether basal toe webbing is present, as it can be important. Only Semipalmated and Western share this characteristic, but it can be very hard to see, and such birds have been called Little or Red-necked merely because the admittedly narrow webbing eluded observation.

It is very important in this group to be aware of a bird's plumage. It's almost axiomatic, especially when trying to pick out rare stints among common ones, to think "plumage first, then species."

Behavior has also been touted as a source of identification, but behavior in all species is at least somewhat varied, usually dependent on substrate (in or out of water, on mud or sand or in vegetation), and either a common or a rare stint in an unusual setting could behave atypically for its species. Comparison with nearby individuals is always one of the best tools for stint identification, or for any other group of birds.

(cont.)

Stints (*cont.*)

In flight, all stints show white wing stripes and a dark rump stripe, but they fall into three groups: (1) Semipalmated, Western, Little, and Red-necked, with the most prominent white wing stripes; (2) Least and Long-toed, usually browner looking and with less prominent wing stripes; and (3) Temminck's, quite brown, like Least, but with flashy white outer tail feathers. Within each group, distinction of species in flight is as much an effort in probabilities as in skill, as purported differences in wing stripes have not been borne out by examination of specimens. Underwing color could be used to distinguish Long-toed.

Flight calls, if given, might be the best guides. Each species has a distinctive call, but the more experience one gains with stints, the more one finds that atypical calls are at least occasionally given, and call should be used as a confirmation of identity, not the sole criterion.

62 Semipalmated Sandpiper

(Calidris pusilla)

This is the most abundant stint in eastern North America, but unlike the widespread Western and Least, it is absent from temperate latitudes in winter. It is drably colored in all three plumages, but nevertheless it provides a common standard for stint identification, and its relatively thick bill is a good field mark.

Size Weight 26 g, length 15 cm (6 in), bill 1.8 cm, tarsus 2.1 cm.

Plumages Bill and legs black. *Breeding adult* varies from plain gray-brown with white supercilium most prominent feature to upperparts regularly or irregularly mottled with dark brown or black. In some individuals, upperparts a bit more reddish brown. Fairly conspicuous supercilium and throat white, narrow loral stripe brown. Breast usually finely striped and spotted with dark brown; sides often with a few dark streaks. Birds in early autumn darker above and more contrasty, as some pale feather edges have faded to white and others have worn off. *Nonbreeding adult* plain gray-brown above, feathers with fine dark shaft streaks; white below, with fine dark streaks across breast. Supercilium white; loral stripe brown, ear coverts like upperparts. *Juvenile* with legs olive at first, darkening to black during fall migration. A variable plumage, typical bird with most feathers fringed with pale buff and tipped with whitish. In some individuals feather centers darker, making fringes more conspicuous and forming distinct scaly pattern. Some birds so pale as to simulate nonbreeding plumage. In many birds overall tones darker, edging not so contrasty, and feathers of upperparts may have dull rufous fringes, in brightest birds almost as highly colored as juvenile

Western. White mantle and scapular lines typical of juvenile calidridines narrow and indistinct or absent in this species. Throat and supercilium white, loral stripe and ear coverts brown. Breast suffused with light tan and streaked with brown on each side, less streaking than in adult. Youngest juveniles with rich buffy breast, fading to whitish during southbound migration.

Identification Similar to other two common North American stints at first glance but easily distinguished from Least with good look, as that species has pale legs, shorter wings, and browner coloration, all of which are evident in every plumage. Much more similar to Western, which differs from Least in all these ways. Western usually has longer bill, much longer in case of female Western, but shortest male Westerns overlap with longest female Semipalmateds, especially in East, where bills of some Semipalmated distinctly longer than in West. Blunter, thicker appearance of bill in most Semipalmateds should distinguish them from shortest-billed Westerns as well as most Leasts, but longest-billed eastern Semipalmateds have thinner bill tip, bill probably indistinguishable from other two species; leg color and plumage characters then important. *Breeding*-plumaged Semipalmateds lack rufous head and back markings and substantial side chevrons that characterize Westerns in that plumage. In *nonbreeding* plumage, long-billed Semipalmated not separable from Western by appearance. Some *juvenile* Semipalmateds and Westerns also very similar, as both species can be quite reddish above. More typical Semipalmateds are buff marked above, rather plainer than Western and often with a scalloped effect, like miniature Baird's Sandpiper. Juvenile Semipalmateds typically have slightly darker crown and ear coverts, thus more contrasty face pattern, than Western. Fortunately, juvenile Westerns seen with longer-billed Semipalmateds in East are usually more advanced in molt, when contrast between rufous upper scapulars and paler lower scapulars and coverts are pronounced in Western. Semipalmated only rarely colored like this. Two Eurasian stints that might occur anywhere in North America, Little and Red-necked, look much like Semipalmated in juvenile and nonbreeding plumages; see those species. Brightest juvenile Semipalmateds surprisingly bright, rarely to Little Stint brightness.

In Flight Typical stint, with moderately conspicuous white wing stripe and dark rump stripe. Stints within a given group (black-legged, pale-legged, Temminck's) probably not distinguishable in flight unless distinctive field marks, usually in breeding plumage, can be seen.

Voice Flight call a short *chert* or *chut* which may be doubled, distinctly lower and shorter than high, rolled call of Least or squeak of Western. Distinctive among common North American stints and also quite different from calls of any of peripheral species. Breeding-ground song a "motorboat" call followed by complex buzzy song.

Behavior One of most abundant shorebirds in eastern North America, with flocks, some of them immense at migration peaks, spread over many habitats. With short bill, feeds by picking and often forages above feeding zone of Western Sandpiper, but probes at times and feeds with Western, especially in shallow freshwater.

Habitat Breeders on moist arctic tundra. Nonbreeders on mudflats, sand beaches, and freshwater lakes and ponds.

Range Breeds from western Alaska all across Arctic to Atlantic coast. Winters from eastern Mexico, southern Florida, and Caribbean islands south on coasts to northern

Chile and southern Brazil. Migrants all across Canada and mostly east of Cascade/Sierra range in USA, rare between British Columbia and Chiapas on Pacific coast. Eastern birds use oceanic route in fall, midcontinent and Atlantic coast in spring. Alaskan and central populations migrate across continent both seasons, Alaskans earlier in spring than other populations.

62.1 Breeding Semipalmated Sandpiper. Short-billed and rather plain stint (stints are smallest sandpipers). Breeding plumage relatively dull, with browns and grays; brightest colors scattered rufous fringes as on crown, a few scapulars, and one tertial here. Upper Coast, Texas, USA, May 2000 (Brian Small).

62.2 Breeding Semipalmated Sandpiper. Birds from eastern part of breeding range with bill longer and more slender (especially in female), approaching Least and Western, but breeding colors still dull. Warren County, Kentucky, USA, Jun 2002 (David Roemer).

62.3 Breeding Semipalmated Sandpiper. Another long-billed eastern bird (presumably also female) but with slightly richer rufous crown. Unusually sparsely marked on breast. Queens, New York, USA, May 2002 (Angus Wilson).

62.4 Breeding Semipalmated Sandpiper. Very worn postbreeding bird with no brown or rufous. Relatively short bill indicates male. Early September but no trace yet of molt into nonbreeding plumage (molt almost complete in Western Sandpipers by that time). Hatteras, North Carolina, USA, Sep 1989 (Henry Lehto).

62.5 Nonbreeding Semipalmated Sandpiper. Nonbreeding birds like dullest breeding birds. This bird has molted most feathers of upperparts on southbound migration, perhaps to winter here. Florida, USA, Sep 1988 (Richard Chandler).

62.6 Juvenile Semipalmated Sandpiper. Typical western juvenile, very dull with no bright rufous anywhere; pattern scalloped, somewhat like juvenile Baird's Sandpiper. Typical Semipalmated head pattern, with complete supercilium and capped effect. Quite short and thick bill distinctive. Ventura, California, USA, Aug 1996 (Brian Small).

62.7 Juvenile Semipalmated Sandpiper. Eastern juveniles can be identical to western, perhaps some from same breeding populations. This set of photos shows how difficult it is to see distinctive toe webbing. Conneaut, Ohio, USA, Sep 2002 (Robert Royse).

62.8 Juvenile Semipalmated Sandpiper. Especially early in season, juveniles can have buffy breast, olive or even yellowish legs. Easily differentiated from Least Sandpiper by duller coloration, paler breast, longer primary projection. Padilla Bay, Washington, USA, Aug 2002 (Stuart MacKay).

62.9 Juvenile Semipalmated Sandpiper. Some eastern juveniles are brighter, with tinge (or more) of rufous on crown and scapulars. Still less rufous than Western, and shorter bill distinctive. This bird shows interrupted, almost split, supercilium, more typical of Little and Red-necked stints. Hatteras, North Carolina, USA, Sep 1989 (Henry Lehto).

62.10 Breeding Semipalmated Sandpiper. Flight pattern typical of stints, with narrow white wing stripe and white sides to rump. Short, relatively thick bill evident here. Fort Myers, Florida, USA, May 2003 (Wayne Richardson).

62.11 Stint wing specimens. Least Sandpiper (bottom) with less distinct wing stripe than Western (top) and Semipalmated. Adult Western Sandpiper, Ocean Shores, Washington, USA, Jun 1990; juvenile Semipalmated Sandpiper, Potholes Reservoir, Washington, USA, Aug 1991; adult Least Sandpiper, Ocean Shores, Washington, USA, Mar 1994.

63 Western Sandpiper

(Calidris mauri)

This relatively large, long-billed stint is the common western stint, surprisingly abundant considering its restricted breeding range. Large numbers move to the southeastern USA into the range of Semipalmated to take its place as the wintering species there.

Size Weight 27 g, length 16.5 cm (6.5 in), bill 2.5 cm, tarsus 2.3 cm.

Plumages Slight dimorphism in bill length (female's longer, usually >2.5 cm, difference usually apparent in comparison). Bill and legs black. *Breeding adult* with much rufous above, concentrated on crown, ear coverts, and feather tracts of back from mantle to scapulars; extent and intensity of coloration variable. Breast streaked, sides marked extensively with black streaks and chevrons. Returning adults in fall grayer and faded, black on back more prominent and patchier, and rufous also more prominent because of exposure of scapular bases. *Nonbreeding adult* virtually identical to Semipalmated Sandpiper, plain gray-brown above and white below, with white supercilium and lightly streaked breast. *Juvenile* with legs olive at first, turning black rapidly. Differs from breeding adult by lack of rufous tones on head and absence of markings on sides. Head pattern like that of nonbreeding adult but darker and slightly more distinct. Mantle feathers fringed with gray-brown or rufous, scapulars and tertials with rufous and white. Coverts gray-brown with buff fringes. Underparts largely unmarked, with streaks on sides of breast and faint buffy wash across it. Typically a white mantle line, more rarely a scapular line. Rusty upper scapulars persist into October. Youngest juveniles with rich buffy breast, fading to whitish during southbound migration.

Identification Largest of stints, but all sufficiently similar that at any distance they are merely stint sized. This species closest to Semipalmated, distinguished most readily by longer bill, with finer, slightly droopy tip. See Semipalmated for great similarities and consistent differences. Juvenile Westerns probably readily identifiable as they invade eastern range of Semipalmated in fall, but wintering birds could be very similar. Westerns distinctly larger than Leasts, less tinted with brown in all plumages, black instead of yellow legs. When Westerns show rufous, contrasts with paler gray-brown, whereas rufous-marked Leasts usually look quite bright reddish brown all over. Both Little and Red-necked have short, fine bills, but when either species turns up as vagrant in North America, important to distinguish from Western. See also casual Spoonbill and Broad-billed sandpipers.

In Flight Typical stint, with moderately conspicuous white wing stripe and dark rump stripe.

Voice Flight call a high-pitched, squeaky *dzheet.* Much less vocal than Least Sandpiper, but calls often heard in big flocks of Westerns. Call clearly monosyllabic, in comparison with Least's rolling trill. Western's call distinctly longer and higher-pitched than Semipalmated's. Breeding-ground song a few long buzzy trills that ascend, then drop at end: *brr-eee brr-eee brr-eee breee-urrrrr.*

Behavior Typical stint, running around rapidly on beaches and mudflats while feeding by both pecking and probing; slows down to probe, which it does more than shorter-billed species. Can be in quite large flocks at roosts and in flight on Pacific coast but does not gather in as immense flocks as Dunlin.

Habitat Breeding on moist lowland tundra. Nonbreeding on mudflats, sand beaches, and freshwater lakes and ponds.

Range Breeds in western Alaska, sparingly on north coast. Winters on coasts from Washington and Delaware south through Middle America and West Indies to Peru and Suriname. Migrants all across continent, abundant in West and scarcest in Northeast; Pacific coast in both seasons, interior and East mostly in fall. More common than Semipalmated in southeastern USA except in spring.

63.1 Breeding male Western Sandpiper. Long-billed stint. Bright rufous on crown, cheeks, and upperparts typical of this species, also dark chevrons and short streaks on white lower breast and sides. Bill length indicates male. Nome, Alaska, USA, Jun 1998 (Brian Small).

63.2 Breeding female Western Sandpiper. Distinctly longer billed than male. Note less bright rufous on upperparts in this individual. Santa Clara County, California, USA, May 2003 (Mike Danzenbaker).

63.3 Nonbreeding Western Sandpiper. Plain gray-brown upperparts, white breast with variable amount of streaking. Bill length indicates male. Few feathers of breeding plumage coming in. Note webbing at base of toes, characteristic of only this and Semipalmated among stints. Sanibel, Florida, USA, Mar 1999 (Brian Small).

63.4 Juvenile Western Sandpiper. Bright rufous on scapulars but not head indicates this plumage. Lacks markings on sides typical of breeding adults. Male by bill length. Ventura, California, USA, Sep 1997 (Brian Small).

63.5 Juvenile Western Sandpiper. Some juveniles show less intense rufous on upper scapulars, but always present. Long bill indicates female. Bolsa Chica Preserve, California, USA, Sep 1995 (Joe Fuhrman).

63.6 Immature Western Sandpiper. Fringed coverts indicate first-year bird. First-year males have shortest bills, could easily be mistaken for Semipalmated; some may be distinguished only by flight calls. Huntington Beach, California, USA, Oct 1995 (Joe Fuhrman).

63.7 Breeding Western Sandpiper. Like Semipalmated Sandpiper (and other stints) in flight; longer bill may be apparent. Rufous of crown and scapulars barely visible here. Fort Myers, Florida, USA, May 2003 (Wayne Richardson).

63.8 Breeding Western Sandpiper. Pattern of underside in flight typical of this species and other black-legged stints. Coyote Point, California, USA, Apr 2000 (Mike Danzenbaker).

64 Red-necked Stint

(*Calidris ruficollis*)

This species is an ecological equivalent of both Eurasian Little Stint and North American Semipalmated Sandpiper, yet it overlaps with both of them on their breeding grounds. Unmistakable in breeding plumage, with its rich rufous head, it is much harder to distinguish from both the common Semipalmated and the rare Little in North America in its other plumages.

Size Weight 26 g, length 15 cm (6 in), bill 1.8 cm, tarsus 2.0 cm.

Plumages Bill and legs black. *Breeding adult* with most of head, neck, mantle, and upper breast rich rufous. Intensity of color variable, some individuals with much white on throat. Rufous of foreparts frosted with white tips early in spring and duller and faded later in summer. Scapulars and tertials heavily marked with black and rufous. Rufous on upper breast bordered behind with scattered dark brown spots that extend a short way down sides. *Nonbreeding adult* light gray-brown above, feathers with conspicuous dark streaks (rarely dark centers); white below, often with fine streaks at side of breast. White supercilium and dark loral stripe. *Juvenile* with legs probably olive at first, as in related species. Relatively dull plumage but considerable variation, with some individuals more brightly marked. Crown and mantle gray-brown to dull rufous with dark stripes. Scapulars often brightly marked, edged with rufous and tipped with white. Tertials typically drab, not especially dark-centered, and fringed with gray-brown to dull buff rather than richer tones of some other species. White mantle and scapular lines usually not well developed. Coverts plain buffy brown. Supercilium typically interrupted by fine streaks above eye. Breast light gray or buffy, with or without fine streaks on sides.

Identification In full *breeding* plumage easily distinguished from all other stints by uniform rufous-orange face, throat, and upper breast. Little Stint also has much rufous on head, but throat and breast white. Both Little and Red-necked vary, and both usually have dark spots on sides of breast, but those spots within orange in Little and typically extend behind it or entirely behind it in Red-necked. Red-necked may show extensive white on throat and breast, probably more often while still in molt, but there are at least scattered rufous patches; also, rufous may be somewhat obscured by paler feather tips when fresh. Red-necked also less strikingly patterned above, typically without whitish mantle and scapular stripes, but some Red-neckeds as bright and contrasty as most Littles. One difference that seems to hold is that Littles always have two or more tertials and often central tail feathers (normally hidden) brightly edged with rufous, whereas Red-necked tertials and tail are entirely dull, occasionally one or more tertials with narrow rufous edges. See Little Stint for differences in other plumages. Orange breeding-plumaged Sanderlings often have been called Red-necked Stints, but they are larger and longer-billed and have entirely streaked breast. Among common stints of North America, Red-necked most like Semipalmated Sandpiper, similar enough in nonbreeding and juvenile plumages to represent major identification challenge. Bill somewhat finer than that of Semipalmated, and wing projections and primary projections slightly longer, but shape differences don't knock your eyes out. In *nonbreeding* plumage, both Red-necked and Little have more extensive dark centers to feathers of upperparts than do Semipalmated and Western, so former pair has slightly more mottled-looking upperparts. In *juvenile* plumage, both Red-necked and Semipalmated quite varied in intensity and shade of markings; differ primarily in head pattern. Red-necked typically has sides of crown more finely streaked then center, producing slightly darker center strip, and supercilium may be slightly interrupted above eye; Semipalmated with crown uniform and simple uninterrupted supercilium, therefore better-defined "cap." Small percentage of each species, unfortunately, patterned in manner typical of other, leaving bill shape and size and presence or absence of toe webbing as one field mark difficult to assess and another difficult to see. See Little Stint, which can be very similar. Perhaps we should be happy that these species don't coexist in many areas, forcing us to record "stint sp." on every field trip.

In Flight Typical stint, with moderately conspicuous white wing stripe and dark rump stripe. All-reddish head diagnostic in breeding birds.

Voice Flight call a high-pitched squeak, rather similar to that of Western Sandpiper but may incorporate trilled element as well; not very audible at a distance. These two species may be difficult to distinguishable vocally, but call of Red-necked quite distinct from those of similar-looking Semipalmated Sandpiper and Little Stint. Breeding-ground song harsh double whistles, given methodically—*correk, correk, correk*.

Behavior Like that of other stints, moving methodically over moist substrates, occasionally running, and picking tiny invertebrates at intervals; as in Semipalmated, probing relatively infrequent. Large flocks where common, usually singles seen in North America.

Habitat Wet tundra and riverbanks for breeding. Mudflats, sandy beaches, lake margins, and marshes for nonbreeding.

Range Breeds from northeastern Siberia to western Alaska. Winters in Southeast Asia and Australasia, south to New Zealand. Migrants along western Pacific shores. Vagrant down Pacific coast to California, more rarely to Atlantic coast and, even more rarely, to interior.

64.1 Breeding Red-necked Stint. Stint with bright rufous face, neck, and upper breast in this plumage. Dark spots extend behind rufous area. Tertials without rufous edgings. Somewhat worn postbreeding individual but still bright. Aomori Prefecture, Japan, Jul 1999 (Mike Danzenbaker).

64.2 Breeding Red-necked Stint. All stints other than Western and Semipalmated sandpipers have unwebbed toes. Aomori Prefecture, Japan, Aug 1999 (Mike Danzenbaker).

64.3 Breeding Red-necked Stint. Some individuals have rufous much less bright, but always in same area and bounded behind by spots. Upperparts can virtually lack rufous. Aomori Prefecture, Japan, Jul 1999 (Mike Danzenbaker).

64.4 Nonbreeding Red-necked Stint. Plain gray-brown above, white below. Streaks at side of breast usually obscure. Bill short, usually more slender than in Semipalmated Sandpiper but just like Little Stint. Primary projection averages longer than in Western and Semipalmated sandpipers, similar to Little Stint. Hyogo Prefecture, Japan, Sep 1994 (Norio Kawano).

64.5 Juvenile Red-necked Stint. No reddish on neck and breast but with patterned upperparts. Often with rufous on upperparts in this plumage but usually not on tertials, as is characteristic of Little Stint. Supercilium usually interrupted and split, often good distinction from Semipalmated Sandpiper, as is long-winged look. Aomori Prefecture, Japan, Aug 1997 (Mike Danzenbaker).

64.6 Juvenile Red-necked Stint. Some juveniles quite rufous above but, again, not so on tertials. Even brightest individuals do not have as distinct mantle and scapular lines as is typical of Little Stint. This individual has capped look somewhat like Semipalmated Sandpiper, indicating this distinction is indicative, not definitive. Miyagi Prefecture, Japan, Sep 2002 (Akihiko Shiraishi).

64.7 Juvenile Red-necked Stint. Some juveniles very dull; this one much like Semipalmated Sandpiper but longer wings. Coverts not very contrasty in this species, better defined with black centers in Little Stint. Hyogo Prefecture, Japan, Sep 2001 (Norio Kawano).

64.8 Juvenile Red-necked Stint. Identical to other black-legged stints in flight. Aomori Prefecture, Japan, Sep 1997 (Mike Danzenbaker).

64.9 Breeding stint wing specimens. Long-toed (bottom) distinctly darker below than all other stints. Little Stint, Yamalo-Nenetskiy, Russia, Jun 1997; Red-necked Stint, Chukotka, Russia, Jul 1992; Long-toed Stint, Magadanskaya, Russia, Jun 1993.

65 Little Stint

(*Calidris minuta*)

This is the most vividly marked stint in breeding and juvenile plumages. Its bright markings set a standard for colorfulness that other stints only occasionally attain. The most common stint in Europe, it is no more than a vagrant to both coasts of North America.

Size Weight 25 g, length 15 cm (6 in), bill 1.8 cm, tarsus 2.1 cm.

Plumages Bill and legs black. *Breeding adult* richly colored above, brightest of all stints in this plumage. Most feathers of crown, cheeks, mantle, scapulars, and tertials dark and fringed with bright rufous, interrupted by white supercilium in some individuals. Usually a distinct pair of white mantle lines. Throat white, contrasting with rufous head; dark streaks and spots on brownish to rufous sides of breast. *Nonbreeding adult* plain gray-brown above, typically with well-defined darker brown feather centers, and white below, with fine streaks on sides of breast. White supercilium and dark loral stripe as in other stints. *Juvenile* with legs olive at first, darkening to black during first autumn. Brightly marked, with conspicuous white mantle and scapular lines. Crown and mantle feathers fringed with rufous. Scapulars and tertials brown, darkening to black toward tips, with bright rufous and white edges. Conspicuous white supercilium and dark loral stripe. Breast varies from white to pale buff, usually with fine streaks on sides. Some individuals with duller pattern.

Identification Rufous-headed *breeding* birds easily distinguishable from all other stints except Red-necked, which has reddish color extending across breast. Little always has white throat and breast; often has reddish back blending with bright head, Red-necked gray back contrasting with bright head. See that species for further distinctions. In *nonbreeding* plumage, Little has fairly large dark feather centers above, making it look spotted. Other black-legged stints look more uniform above, with fine dark shaft streaks rather than blotches, although Red-necked can be patterned much like Little. *Juvenile* Little usually most brightly marked of stints, but nevertheless some Littles dull, and brightest juveniles of each stint species very bright indeed. If bill length can be used to separate Western from Little (but even those two species have been confused), then latter most like Red-necked and Semipalmated in juvenile plumage. Contrasty rufous-edged tertials of Little present good field mark, as other short-billed, black-legged stints have rather plain tertials. Little almost always brighter than Semipalmated, much more likely to have white mantle and scapular lines, also more likely than Red-necked. Supercilium usually more obviously split in Little than Red-necked. Dark central markings of lower scapulars usually large and bold in Little, narrower and less conspicuous in Red-necked and Semipalmated, often wider toward tip or anchor shaped. Scapulars in Little usually have lengthy white fringes, those in Red-necked with less white, typically looking tipped with white. Stripes on sides of breast a bit more conspicuous than in Red-necked. In Japan, where sometimes seen together, Little distinguished by slightly longer wings than Red-necked, projecting slightly farther beyond tail. Some stints will defy definition! Flight calls of these similar stints may aid in distinguishing them, but bear in mind variation around "typical" call of each species. Although juvenile Least Sandpipers have differently colored legs, and thus should be easily distinguished, they are often as bright and contrasty as Little Stints, and as observers may not note leg color immediately or it may be obscured, that species must be eliminated when Little Stint being considered. Least has shorter wings and primary projections and, usually, fine streaks and brownish wash all across breast.

In Flight This species has moderately conspicuous white wing stripe and dark rump stripe, like Red-necked Stint and Western and Semipalmated sandpipers.

Voice Flight call a sharp single note, squeaky but shorter than Western Sandpiper's and Red-necked Stint's calls. Has some qualities of calls of Sanderling and Red-necked Phalarope but distinct from both. May give trilled note, more similar to calls of other stints.

Behavior Hyperactive, like most small sandpipers. Picks at surface prey and probes shallowly. Forms large flocks where common but likely to be seen singly in North America.

Habitat Tundra for breeding. Mudflats, beaches, and lake margins for nonbreeding.

Range Breeds across arctic Eurasia from northern Scandinavia well east in Siberia but not to Pacific. Winters in sub-Saharan Africa and on Indian subcontinent. Migrants widely on fresh and salt waters, rarely east to Bering Sea islands. Vagrant elsewhere in North America, on both coasts and much more rarely inland, more often in fall than spring; once in Baja California and twice in West Indies.

65.1 Breeding Little Stint. Black-legged stint with much rufous above, including on ear coverts and tertials. Heavily spotted on upper breast but lacks streaks and chevrons on sides characteristic of Western Sandpiper, also shorter bill. Salo, Finland, May 1993 (Henry Lehto).

65.2 Breeding Little Stint. Some individuals very bright, with mostly reddish head but always white throat. Often rufous on sides of breast, heavily spotted but spots do not extend behind rufous (as in Red-necked Stint). Cyprus, May 2003 (Mike Danzenbaker).

65.3 Breeding Little Stint. Upperparts may be vividly striped in this species, here with black-bordered white mantle lines and less well defined upper scapular lines. Toes unwebbed. Odiorne Point State Park, New Hampshire, USA, Aug 2003 (Glen Tepke).

65.4 Breeding Little Stint. Some individuals show no rufous on head; perhaps molt into breeding not complete. Lesbos, Apr 2002 (Verbanck Koen).

65.5 Breeding Little Stint. Worn late-summer individual with few feathers of nonbreeding plumage coming in. Unusual in showing no trace of rufous on tertials, perhaps much of it worn off. Ferrara, Italy, Aug 2001 (Fabio Ballanti).

65.6 Nonbreeding Little Stint. Plain in this plumage like other black-legged stints, but feathers of upperparts have more distinct dark centers, giving mottled look. Very like Semipalmated Sandpiper and Red-necked Stint (see text). Sardinia, Jan 1995 (Massimo Piacentino).

65.7 Juvenile Little Stint. Typical strongly patterned juvenile with well-defined mantle and upper scapular lines, much rufous above, dark-centered coverts. Most have split supercilium, well-defined short dark streaks on sides of breast. Miyagi Prefecture, Japan, Sep 2002 (Akihiko Shiraishi).

65.8 Juvenile Little Stint. Darker and more uniformly colored individual with strong head pattern; rufous tertial edging typical of species and plumage. Long primary projection typical of Little and Red-necked. Aichi Prefecture, Japan, Sep 1987 (Shinji Koyama).

65.9 Immature Little Stint. Like nonbreeding, with dark-centered feathers of upperparts, but very worn juvenile coverts (still with some rufous) and tertials evident. Aichi Prefecture, Japan, Feb 2003 (Koichi Tada).

65.10 Juvenile Little Stint. Like other black-legged stints in flight. Typical mantle lines and split supercilium visible. Cleveland, England, Sep 1995 (Wayne Richardson).

66 Temminck's Stint

(Calidris temminckii)

Temminck's is the most distinct stint, with its scalloped juvenile plumage, plain nonbreeding plumage, long tail, rather solitary behavior, and towering flight. Although probably the least common stint on a worldwide basis, nowhere in large numbers, it tends to stay out of flocks and thus remains detectable.

Size Weight 24 g, length 14 cm (5.5 in), bill 1.7 cm, tarsus 1.8 cm.

Plumages Bill black; legs yellowish to greenish, usually pale but may be fairly dark. *Breeding adult* gray-brown and rather plain above, with scattered irregular markings of black, buff, and rufous. These markings most distinct on scapulars, where some may be chevron shaped. White eyering conspicuous against dull facial pattern. Breast washed with same color as back and finely streaked, looking plain from distance. Autumn adults quite dull; reddish markings of spring subdued and mostly restricted to scapulars, breast with small spots. *Nonbreeding adult* has plain gray-brown back and breast, relieved only by faintest of dark streaks on upperparts. Face plain, narrow white eyering conspicuous, gray-brown loral stripe inconspicuous. Breast washed with light gray-brown, unstreaked or with faintly indicated streaks. *Juvenile* colored like nonbreeding adult but with most feathers of upperparts fringed with whitish to buff, with narrow brown subterminal line. These markings like those of juvenile Surfbird and Red Knot and different from dark-centered feathers of other stints. Some with subterminal fringes on scapulars very broad, producing barred effect. Pale fringes may wear off by late September, but scalloped pattern produced by dark fringes more persistent.

Identification Seems short legged, pointed behind, something like small version of Baird's and White-rumped sandpipers; long tail often extends beyond wingtips. Bill quite small and slender. If yellow legs clearly seen, then only two other yellow-legged stints similar. From both common Least and rare Long-toed, distinguished by plain, unmarked brown breast (some individuals of other species with streaks obscure). In breeding plumage plainer than other stints, often with only scattered dark markings above. In nonbreeding plumage, little indication of darker markings on gray-brown upperparts and breast, actually reminiscent of Spotted and Common sandpipers but lacks white mark in front of wing of those species. In juvenile plumage, only stint with pale feather edgings that produce scalloped look. No trace of mantle lines in any plumage.

In Flight Inconspicuous white wing stripe like several other stints, but tail quite different, relatively long and brown with no distinct dark stripe down central feathers, outer feathers shining white. White outer tail feathers should provide good field mark in flight but aren't always easy to see in flushed bird. Temminck's more likely than other stints to "tower" straight up like snipe or Solitary Sandpiper.

Voice Flight call a short, dry, cricketlike trill, often given in series. Usually vocal when flushed, recalling Solitary Sandpiper or snipes with similar behavior.

Behavior Usually single or in small flocks even where common. Feeds high on mudflats or in marsh vegetation like Least and Long-toed, often creeping along with flexed legs like Least. Tends to occur at small water bodies.

Habitat Wet tundra for breeding. Sheltered estuarine mudflats and freshwater marshes and lake margins for nonbreeding.

Range Breeds across Eurasia from Scandinavia to Chukotski Peninsula. Winters around Mediterranean, in equatorial part of Africa, and across Southeast Asia mainland. Migrants mostly on freshwater across Europe and Asia, rarely east to Bering Sea islands. Vagrant to southern British Columbia in fall.

66.1 Breeding Temminck's Stint. Yellow-legged stint with rather attenuate shape and plain plumage. Variably patterned with black blotches and chevrons and some reddish fringes; breast well streaked. Cyprus, May 2003 (Mike Danzenbaker).

66.2 Breeding Temminck's Stint. Worn postbreeding, with black markings prominent, pale fringes worn off. Cyprus, Aug 2002 (Mike Danzenbaker).

66.3 Nonbreeding Temminck's Stint. Distinctive in plainness, including all-gray-brown breast. White eyering prominent against plain face with poorly defined pattern. Bill at small end for stint. White outer tail feather apparent. Cyprus, Nov 2002 (Mike Danzenbaker).

66.4 Juvenile Temminck's Stint. Unique among stints in pattern, much like nonbreeding but with fine dark subterminal lines and pale fringes on feathers of upperparts. Tail long, typically extending beyond wings. Kyoto Prefecture, Japan, Sep 2001 (Norio Kawano).

66.5 Immature Temminck's Stint. Like nonbreeding but with retained juvenile coverts. Some individuals with essentially patternless head, unlike any other stints. Hyogo Prefecture, Japan, Oct 2001 (Norio Kawano).

66.6 Breeding Temminck's Stint. Upper- and underwing patterns as in most other stints. White outer tail feathers visible. Varanger, Norway, Jun 1998 (Mike Lane).

66.7 Breeding Temminck's Stint. Wing stripes as in other stints but different tail pattern. Tail long, with pure white outer feathers. Yyteri, Finland, Aug 1991 (Markku Huhta-Koivisto).

67 Long-toed Stint

(Calidris subminuta)

This stint can almost be called elegant because of its long legs and upright posture, but it still looks too much like other species in its various plumages to be easily recognized. From its limited breeding range in Siberia, small numbers move into North America during migration each year.

Size Weight 24± g, length 14.5 cm (6 in), bill 1.8 cm, tarsus 2.2 cm.

Plumages Bill black, often with pale base of lower mandible; legs yellowish, rarely yellowish brown or greenish yellow. *Breeding adult* in fresh plumage brightly patterned above, with extensive rich rufous edges on crown, mantle, scapulars, and tertials producing predominantly striped pattern. Some scapulars tipped with white, most tertials entirely fringed with rufous. Prominent white supercilium, dark loral stripe usually expanded into smudge before eye, reddish ear coverts. Breast finely streaked with black and washed with buff in some individuals. Early fall adults look much darker above, like Least Sandpipers, when large proportion of rufous fringes worn off. *Nonbreeding adult* overall plain gray-brown; crown and hindneck streaked with blackish and feathers of mantle, scapulars, and some tertials conspicuously centered with dark brown, causing characteristic blotched appearance. Prominent white supercilium and dark loral stripe. Breast gray-brown, varying from uniformly colored to finely streaked. *Juvenile* marked rather like breeding adult but much brighter above than any worn adult with which it might occur. Feather edgings vary from wide and bright rufous to narrower and duller, but less likely than adult to show buff breast or rufous cheeks. Usually vivid white mantle lines and less well defined scapular lines. All feathers of upperparts fringed with rufous and/or buff. White supercilium distinct, dark loral stripe less so than in breeding adult and not usually expanded into smudge in front of eye. Breast light gray-brown with brown streaks. Reminiscent of juvenile Sharp-tailed Sandpiper.

Identification With overall brown to reddish brown coloration, yellow legs, and relatively short wings, most like Least Sandpiper. Differs from that species by slightly larger size and slightly longer legs—toes as well as tarsus—which make bird stand a bit higher. In Long-toed, toes slightly longer than bill, in Least about same length; not easy to use as field mark. Posture more upright and alert than Least on average, but either species can stand alertly while being approached by predator (or birder) or can hunch over while foraging. Lower mandible usually pale at extreme base; Least shows this only rarely. In both breeding and juvenile plumages, differs from Least in back pattern, mantle looking striped rather than scalloped; one of best field marks. In these plumages, Long-toed also shows pale gray or whitish breast with fine streaks (browner/buffier in Least, with coarser streaks). A showy stint in *breeding* plumage, with reddish cap (brown in Least) and tertials broadly fringed with bright rufous (narrowly fringed and typically with duller rufous in Least). This difference persists into late summer, when fringes on Least may be worn off completely. *Nonbreeding* plumage very like that of Least Sandpiper but feathers of upperparts with larger black centers, thus looking blotched. Head pattern of the two differs in *juvenile* plumage, difference not as distinct in breeding plumage. Long-toed has conspicuous white supercilium bordered in front by dark forehead and rounded at anterior end, whereas white supercilium of Least narrows at anterior end and usually meets fellow from other side as pale forehead. Supercilium often split in Long-toed, not in Least. Juveniles differ further in Long-toed having buffy scapular line (white in Least) and whitish-edged wing coverts (buff edged in Least, contrasting less with rufous-fringed scapulars). Recall that other juvenile stints can be strikingly marked; short primary projections, fine breast streaks, and yellow legs distinguish Long-toed from brightest species, Little and Western.

In Flight This species and Least Sandpiper are like other stints but typically browner above, wing stripe quite inconspicuous. Should be distinguishable with good look in flight from black-legged stints, which are usually paler in most plumages and have more conspicuous wing stripes. This species has darkest underwing, lacking expanse of white coverts typical of other stints, but unlikely to be seen easily. Only stint in which toe tips extend beyond tail in flight; may be excellent field mark.

Voice Flight call a rolling *prrrp*, distinctly lower pitched than that of Least. Bear in mind that Least occasionally gives a lower-pitched call, so call not enough in itself for definitive identification.

Behavior Feeds among dense low vegetation, on mud, or in shallow water, picking prey from surface. Stance upright, only stint ever likened to *Tringa* sandpipers in posture; has been likened to Wood Sandpiper.

Habitat Wet tundra for breeding. Mudflats, lake shores, and marshes for nonbreeding.

Range Breeds in relatively restricted region of eastern Russia, within a few hundred miles of Pacific Ocean. Winters in Southeast Asia and Australia. Migrants east regularly to Bering Sea islands. Vagrant down American Pacific coast to Oregon and California.

67.1 Breeding Long-toed Stint. Yellow-legged stint with no primary projection, most similar to Least Sandpiper but with rather long legs. This plumage with reddish on cap and upperparts. Breast pale buff with fine streaks (browner in Least). Kyoto Prefecture, Japan, Jul 2000 (Norio Kawano).

67.2 Breeding Long-toed Stint. Head pattern distinctive, with white supercilium not extending to bill and dark vertical smudge in front of eye. Unusually plain upperparts in this bird, almost a mixture of breeding and nonbreeding plumage; perhaps immature. Tatsuta, Aichi Prefecture, Japan, Apr 2003 (Koichi Tada).

67.3 Breeding Long-toed Stint. Worn postbreeding individual, with much of rufous edging worn off. Aomori Prefecture, Japan, Sep 1998 (Mike Danzenbaker).

67.4 Nonbreeding Long-toed Stint. Plain yellow-legged stint, similar to Least but feathers of upperparts with dark centers, giving blotched effect. Breast averages paler; supercilium often split and not reaching bill (forehead dark). Note long toes (show beyond tail tip in flight). Okinawa Prefecture, Japan, Jan 2000 (Shinji Koyama).

67.5 Juvenile Long-toed Stint. Somewhat like breeding adult but even more brightly marked, with strong white mantle lines. Distinguished from Least Sandpiper by dark forehead, split supercilium, striped mantle, wide rufous edges on scapulars and tertials contrasting with duller coverts, better-defined breast streaks. Base of lower mandible often pale in this species, not in Least. Aomori Prefecture, Japan, Sep 1999 (Mike Danzenbaker).

67.6 Juvenile Long-toed Stint. All characteristics visible on this individual, including long tarsus and middle toe (longer than bill). Kyoto Prefecture, Japan, Aug 2000 (Norio Kawano).

67.7 Juvenile Long-toed Stint. Much duller juvenile, still shows striped mantle, supercilium ending before bill, contrast between rufous scapulars and gray coverts. Salinas, California, USA, Aug 1988 (Larry Sansone).

68 Least Sandpiper

(Calidris minutilla)

The smallest shorebird, this tame species is common across North America but never seen in the numbers of the other two common stints. Not always easy to see as it moves slowly through marsh vegetation, it feeds with other stint species but usually separates into flocks of its own species for flying and roosting. Although varying greatly in overall brightness, Least Sandpipers look browner than the other common North American stints in any plumage.

Size Weight 21 g, length 14 cm (5.5 in), bill 1.9 cm, tarsus 1.9 cm.

Plumages Bill black, legs dull to bright yellowish. *Breeding adult* light brown and black above, with fringes buff but varying to rufous, tipped with whitish in some individuals. Whitish supercilium and dark loral stripe fairly conspicuous. Breast with conspicuous dark streaks, demarcation between it and white belly marked. Returning fall migrants have light edges variably worn off, majority of birds looking black backed because of this, although some have buff fringes faded to whitish, retaining conspicuous pattern. *Nonbreeding adult* light brown on back and breast, slightly darker feather centers lending a faintly mottled appearance. Some birds look just about plain, others more conspicuously streaked. White supercilium only moderately well defined. *Juvenile* marked above with same colors as breeding adult but generally brighter. Crown and mantle feathers fringed with reddish brown. Scapulars variably marked with reddish on bases, edged with buff, and tipped with white. White mantle lines conspicuous, those on scapulars slightly less so. Tertials narrowly fringed with rufous. White supercilium and dark loral stripe conspicuous. Breast varies from light brown to pale buff; streaks finer and less distinct than in adults, in some individuals lacking from center of breast.

Identification Distinguished from other two common North American stints by smaller size, yellow legs, browner coloration in all plumages. Breast distinctly brownish, same color as upperparts, in contrast with pale breasts of Semipalmated and Western. Bill shorter and finer than Western's, finer than Semipalmated's, in some individuals slightly droopy. Long-toed Stint, rare in North America, is most similar; see that species. Juvenile Least colored much like juvenile Pectoral but much smaller, shorter-winged.

In Flight Like other stints but typically browner above, wing stripe quite inconspicuous. See Long-toed Stint. When Least and Western or Least and Semipalmated sandpipers occasionally flock together in flight, Least stands out as darker and smaller.

Voice Flight call a high-pitched, clearly two-syllabled or slightly rolled *kree-eeet.* Quite different from Western's squeak and Semipalmated's short single note; in fact, not really like any other shorebird. Breeding-ground song a rhythmically repeated *breee breee breee* and more complex song of single and broken notes.

Behavior Our smallest sandpiper, reminiscent of a feathered mouse when foraging in marsh vegetation with short steps, hunched body, and downward-directed bill. Prefers upper edge of mudflats with scattered to dense vegetation, a drier zone than that frequented by other North American stints, but will wander far away from it. Almost never on open sandy beaches. Feeds by picking and shallow probing. Usually collects into roosts of its own species rather than mixing with other stints.

Habitat Sedge meadows and coastal wetlands in boreal forest and moist tundra for breeding. Fresh and salt marshes and freshwater lake and pond shores for nonbreeding. In winter, more common inland than other stints.

Range Breeds across boreal forest at subarctic latitudes from western Alaska to Newfoundland, south to Queen Charlotte Islands and James Bay. Winters across southern USA and up Pacific coast to Washington and Atlantic coast to North Carolina; also throughout mainland tropics and West Indies, southern extent of winter range northern Chile, Bolivia, and central Brazil. Migrants throughout continent.

68.1 Breeding Least Sandpiper. Yellow-legged stint with short wings (no primary projection). Black feather centers with brown to rufous edges above, heavily streaked brownish breast. Desert Center, California, USA, May 1995 (Joe Fuhrman).

68.2 Breeding Least Sandpiper. Richly colored individual but lacks white lines on upperparts that similarly colored Little Stint would show, and of course has yellow legs. White forehead and unsplit supercilium should distinguish from Long-toed Stint, which also looks longer legged than this. Ventura, California, USA, May 1997 (Brian Small).

68.3 Breeding Least Sandpiper. Always brownish wash across breast. Most have white forehead, but like so many other field marks, not totally consistent. Presence or absence of toe webbing in different stint species is consistent, however. Queens, New York, USA, May 2002 (Angus Wilson).

68.4 Breeding Least Sandpiper. Worn postbreeding birds can look almost black and white, with all brown and rufous faded or worn off. Dramatic difference between this bird and the ones shown above, all in the same plumage, illustrates the problems field guides have always had when using one illustration for a bird, as well as the reason why shorebird identification has always been considered difficult. Seattle, Washington, USA, Jul 1983 (Dennis Paulson).

68.5 Nonbreeding Least Sandpiper. Plain brown back and breast in this plumage; feathers of upperparts either uniform or with some dark centers. San Diego, California, USA, Feb 2003 (Brian Small).

68.6 Juvenile Least Sandpiper. Quite bright individual, with white lines on mantle, upper scapulars, and lower scapulars. At most extreme, rich rufous edgings on all feathers of upperparts (not on coverts in other stints, with exception of Little). Loral stripe complete, white supercilium extends to forehead (but note hint of split supercilium, not typical of Least). Kent, Washington, USA, Aug 2003 (Stuart MacKay).

68.7 Juvenile Least Sandpiper. Alert Least Sandpiper can present appearance more typical of Long-toed Stint. Identified as Least from scalloped instead of striped mantle, white forehead (and of course location). Cape May, New Jersey, USA, Sep 1991 (Erik Breden).

68.8 Juvenile Least Sandpiper. Typical juvenile, beginning molt into first-winter plumage much like adult. Scapulars almost always first feathers to molt in fall. Note coverts buff rather than rufous edged, again showing inadequacy of most single field marks in stint identification. Ventura, California, USA, Sep 1997 (Brian Small).

68.9 Breeding Least Sandpiper. Quite brownish stint with rather inconspicuous wing stripe. Saint Petersburg, Florida, USA, May 2003 (Wayne Richardson).

69 White-rumped Sandpiper

(*Calidris fuscicollis*)

This is one of the shorebirds that are common on the North Atlantic coast in fall and in midcontinent in spring. A surprisingly late fall migrant, it is still common in North America when most other species heading for South America have disappeared southward. It would be intellectually satisfying to understand why this small *Calidris* has a white rump, unlike its close relatives.

Size Weight 44 g, length 18 cm (7 in), bill 2.9 cm, tarsus 2.4 cm.

Plumages Bill black, extreme base of lower mandible dull reddish; legs black. *Breeding adult* gray-brown above with paler brown fringes. Ear coverts, crown, mantle, and upper scapulars may be lightly tinged with rufous. Often faintly indicated light buff mantle and scapular lines. White supercilium fairly conspicuous. Breast with fine black stripes that continue prominently down sides, where mixed with or largely replaced by chevrons. Adults in early fall look darker and more blotchy above, may lack any indication of rufous. *Nonbreeding adult* with upperparts gray-brown with dark feather centers varying from fine shaft streaks to prominent oval markings. White supercilium conspicuous against evenly colored head, breast, and upperparts. Breast streaks less conspicuous than in breeding plumage, sides sparsely streaked if at all. *Juvenile* with upperparts a bit brighter and browner than in breeding adult, often with rufous edges to crown, ear coverts, mantle, scapulars (especially upper ones), and tertials. White supercilium, dark loral stripe, and white mantle and scapular lines prominent. Cheeks, throat, and breast more faintly streaked than in breeding adults, sides rarely so; overall impression a bird more heavily marked above and much more lightly marked below than breeding adult.

Identification Same size and shape as Baird's but differs in having pale base of lower mandible, rufous markings in breeding and juvenile plumages. Breeding birds with streaks along sides. Colored rather like Western in breeding and juvenile plumages but differs in size, more attenuated look, and not quite so much rufous on upperparts. Nonbreeding birds very much like Baird's in being plain and uniformly colored, but gray (Baird's a bit browner), and supercilium more prominent behind eye than in Baird's. Often some streaks on sides in nonbreeding White-rumped, never in Baird's. White rump patch usually impossible to see unless bird opens wings.

In Flight White wing stripe inconspicuous, rump white and tail gray. Distinguished from similar-sized calidridines by white rump, lacking black line characteristic of group. Somewhat like much rarer Curlew Sandpiper but that species slightly larger, with more extensive white rump patch and longer legs, toes of which project distinctly beyond tail; bill-length difference should be evident.

Voice Flight call a mouselike squeak, even thinner and higher pitched than similar call of Western Sandpiper, quite distinctive. Flushed birds likely to call. Aerial display on breeding ground accompanied by fishing-reel-like buzzing and oinks like small pig.

Behavior Moves slowly over mud or through shallow water, makes a few quick probes, and moves again. Often in large flocks and roosting with other species.

Habitat Moist tundra for breeding. Beaches, mudflats, and lake shores for nonbreeding.

Range Breeds in Arctic from northern Alaska across northern Canada to Baffin Island. Winters in southern South America, rarely north to West Indies. Migrants primarily in midcontinent in spring, off Atlantic coast in fall, juveniles surprisingly late. Many down east coast of Middle America and through Caribbean in fall, but wanders widely in both seasons. Very rare west of Rocky Mountains, only handful of records on Pacific coast of USA, none from Mexico, few from countries farther south.

69.1 Breeding White-rumped Sandpiper. Small, long-winged, black-legged calidridine. Rather dull in this plumage but with strongly streaked breast and, often, sides. Atlantic County, New Jersey, USA, May 1995 (Mike Danzenbaker).

69.2 Breeding White-rumped Sandpiper. Dullest version of breeding plumage, but note hint of rufous on anterior scapulars. Nonbreeding looks much like this. Pale base of lower mandible distinctive but not always apparent. McAllen, Texas, USA, May 1998 (Brian Small).

69.3 Breeding White-rumped Sandpiper. White rump usually difficult to see, visible only when wings lifted. Churchill, Manitoba, Canada, Jun 1982 (Dennis Paulson).

69.4 Breeding White-rumped Sandpiper. Most of plumage retained dull breeding, even though late in season, but very drab as in nonbreeding plumage. Fresh (rounded) pale scapulars contrast with worn (pointed) dark-centered ones. Breast streaks less distinct and often obscure in this plumage. Conneaut, Ohio, USA, Oct 2001 (Robert Royse).

69.5 Juvenile White-rumped Sandpiper. More brightly marked than other two plumages, usually with rufous on crown, mantle, scapulars, and tertials, and often with white mantle and scapular lines. Colored somewhat like Western Sandpiper but more heavily streaked breast and longer wings. Note pale base of lower mandible, characteristic when present. Locality, date uncertain (Brian Small).

69.6 Juvenile White-rumped Sandpiper. Much duller example of this plumage (very late in season), almost Baird's-like in its scaly appearance, but basically gray rather than buffy and with hint of reddish on upper scapulars. Presquile Provincial Park, Ontario, Canada, Nov 1989 (Tadao Shimba).

69.7 Breeding White-rumped Sandpiper. Small calidridine with white rump; only similar species Curlew Sandpiper, with longer bill and longer legs (toes readily visible beyond tail tip) as well as more extensive rump patch. Brigantine National Wildlife Refuge, New Jersey, USA, May 1995 (Mike Danzenbaker).

70 Baird's Sandpiper

(*Calidris bairdii*)

Long winged and thus rakish looking, Baird's Sandpiper is always a special sight moving among the smaller stints. Adults are localized migrants, whereas juveniles occur all across the continent. This is one of the few shorebirds that seem to prefer mountain lakes during southbound migration.

Size Weight 38 g, length 18 cm (7 in), bill 2.2 cm, tarsus 2.2 cm.

Plumages Bill and legs black. *Breeding adult* with upperparts mottled black, brown, and buff, tertials quite plain. Faintly indicated paler supercilium and darker loral stripe. Breast light brown, streaked with dark brown. In fresh-plumaged spring migrants, feathers evenly and widely margined with warm, almost buffy brown; paleness prevails. In worn individuals in early fall, pale fringes faded and narrower, allowing dark centers to dominate and present more irregularly blotched appearance. *Nonbreeding adult* much plainer than breeding-plumaged adult, with overall brownish cast and little evidence of dark feather centers; supercilium and loral stripe slightly more conspicuous. Breast streaks finer and less conspicuous than in breeding birds. Rarely seen in North America. *Juvenile* with legs olive at first, darkening quickly to black. A smoothly and evenly marked plumage, quite distinct from blotchy-looking fall adults. All feathers of upperparts fringed with light brown to buffy white. Fringes on scapulars, coverts, and tertials produce more distinctly scaled appearance than in most other *Calidris* juveniles. No trace of mantle or scapular lines. Breast contrasting with belly, light buffy brown with fine and indistinct dark streaks less conspicuous than stripes of breeding-plumaged adults. Buff fringes begin to wear off during migration.

Identification Small sandpiper with long wings, most like White-rumped in size and shape but lacking reddish markings of that species in breeding and juvenile plumages. Breeding adults lack black markings on sides characteristic of White-rumped. Juveniles with scaly rather than striped back. Nonbreeding adults very like White-rumped but browner, with slightly more distinct feather markings; rump should be seen for definitive identification. Breeding adults colored much like Semipalmated but larger, longer-winged, thinner-billed. Note that some juvenile Semipalmateds very brightly marked and somewhat scalloped, could be mistaken for Baird's. Juvenile Baird's with heavily streaked breast on buffy ground color fairly sharply set off from white belly could be mistaken for juvenile Pectoral, but black legs and scaly back distinctive. See also Sanderling.

In Flight Inconspicuous white wing stripe and dark tail stripe like many other small *Calidris*. Could be confused with smaller stints but less conspicuous wing stripe than black-legged species, more like Least. Very similar to slightly larger and darker Pectoral, with its slightly less conspicuous wing stripes, but frustratingly, even flight calls similar. Experience, good size judgment, and a good ear are your only hope.

Voice Flight call a rolling trill, similar to Pectoral's but a bit higher pitched and more musical sounding. Trill slightly slower, individual notes better separated from one another, intermediate between Pectoral and Least. Calls frequently in flight. Breeding-ground display song a buzzy, musical *bur-rrreee bur-rrreee,* may be extended into a longer trill.

Behavior In migration more likely to forage on relatively dry substrates than other *Calidris* sandpipers, often at upper edges of beaches, but also perfectly likely to be on mudflats with Pectoral, Semipalmated, and others, even feeding out in water of freshwater lakes.

Habitat Dry tundra for breeding. Sand beaches, mudflats, and freshwater lake margins in migration and winter.

Range Breeds across Arctic from western Alaska to Baffin Island and northern Greenland. Winters in southern South America, both in mountains and on coasts. Migrants primarily up and down midcontinent, including Mexican Plateau; also Rocky Mountains in fall. Adults rare on Pacific coast and very rare on Atlantic coast, juveniles commonly to both coasts. Very rare migrant in West Indies and northern South America east of Andes.

70.1 Breeding Baird's Sandpiper. Black-legged, long-winged, small calidridine with rather dull brown and black upperparts. This individual already well worn from arctic sojourn. Patterned most like Semipalmated but long wings diagnostic. Toes unwebbed. Prudhoe Bay, Alaska, USA, Jun 1993 (Keith Brady).

70.2 Nonbreeding Baird's Sandpiper. Very plain in this plumage but buffy wash, as in juvenile, distinguishes it from similar-shaped but quite gray-brown White-rumped. Compare size of scapulars with juvenile. Tierra del Fuego, Chile, Dec 1998 (Hanne and Jens Eriksen/VIREO).

70.3 Juvenile Baird's Sandpiper. Quite buffy in this plumage, feathers of upperparts all fringed with whitish buff, giving strong scaly effect. Long wings and primary projection distinguish from stints. Only similarly shaped species is differently colored White-rumped. Ventura County, California, USA, Sep 1998 (Brian Small).

70.4 Juvenile Baird's Sandpiper. Breast heavily streaked but not quite so contrasty as in Pectoral and lower edge slightly indented rather than coming to a point as in Pectoral. Black legs also distinguish from Pectoral and Least. Crittenden Marsh, California, USA, Aug 2002 (Peter La Tourrette).

70.5 Baird's Sandpiper. Typical small calidridine flight pattern, much like Least and Pectoral in having inconspicuous wing stripe. Jackson Reservoir, Colorado, USA, Aug 2003 (Bill Schmoker).

71 Pectoral Sandpiper

(*Calidris melanotos*)

This sandpiper is special because of its substantial size dimorphism, a consequence of males competing to mate promiscuously with as many females as possible. The species prefers to forage in marshes, and an observer who alarms a flock will see heads popping up all over the place from the low vegetation.

Size Male weight 96 g, length 22 cm (8.5 in), bill 3.0 cm, tarsus 3.0 cm; female weight 65 g, length 20 cm (8 in), bill 2.7 cm, tarsus 2.7 cm.

Plumages Moderate dimorphism in size (male distinctly larger). Bill black or dark brown, often brown, yellow, or dull greenish on basal half; legs dull yellow to greenish yellow. *Breeding adult* overall fairly bright, buff to reddish above, streaked with darker. Whitish supercilium and darker loral stripe inconspicuous. Foreneck and breast light brown with blackish stripes, augmented on lower breast by transverse markings. Lower breast of displaying male looks very dark. Bright fringes on upperparts faded to dull light brown in most adults in fall, but some remain almost as bright as juveniles. *Nonbreeding adult* dull version of breeding plumage, with most pale areas light brown. Little hint of bright buff or reddish except, in some individuals, on crown and on few scapulars and/or tertials. These birds not very different looking from worn breeding adults. *Juvenile* brightest plumage of species, patterned basically like adult but with bright rufous edges to crown feathers, scapulars, and tertials in fresh plumage. White lines on both mantle and upper scapulars distinct, and in some a ghost of third line bordering lower scapulars. Moderately distinct whitish supercilium and dark loral stripe. Breast buffier, stripes finer than in adults. Some juveniles duller, with rufous edges darker and reduced in width, and even fewer virtually lack rufous. At close range, worn tertials of autumn adults usually obvious and allow easy distinction from juveniles.

Identification Brownest of small sandpipers clearly larger than stints, strong contrast of dark-striped buffy brown breast and white belly distinctive. Distinguished from other streak-breasted species by brown of breast ending in point in middle. Juveniles vividly striped above, pattern as distinct as in any striped juvenile calidridine but not in realm of snipe striping; also with much reddish above in this plumage. Only other calidridines in this size range with yellowish legs are Ruff and Buff-breasted, neither with strong breast/belly demarcation or well-striped back, and both would be called buffy rather than reddish. Adults not as brightly marked as juveniles but breast pattern still distinguishes them from other species. See very similar but much rarer Sharp-tailed Sandpiper.

Hybrids I must mention the enigmatic "Cox's Sandpiper" (*Calidris paramelanotos*) here. Considered a hybrid between Pectoral and Curlew sandpipers by most ornithologists (based on analysis of DNA and proteins from Australian birds, as well as morphological intermediacy), it is a very rare but regular migrant to southeastern Australia, presumably from Siberia (see Figs. I.7 and I.8 in the Introduction). A bird identified as this "species" was seen in Massachusetts in September 1987, and one may occur again in North America. The 1987 bird, a juvenile, looked much like a Pectoral Sandpiper except for a less prominently streaked breast, gray-brown wing coverts without rufous edgings, longer yellowish legs, and—most distinctive—a considerably longer, slender and slightly droopy, all-black bill rather like that of Curlew Sandpiper. In addition, an unusual adult specimen collected in New York in May 1833 was named "Cooper's Sandpiper" (*Calidris cooperi*), and this is also surely a hybrid between two *Calidris* species. However, at least one of its parents was probably different from those producing "Cox's." Birds corresponding to "Cooper's" also have been seen in Australia.

In Flight Quite dark, brown except for white belly; white wing stripe inconspicuous, dark tail stripe conspicuous. Most like Baird's in flight, also something like Ruff and Buff-breasted but less patterned above than former, more patterned above than latter and with strong breast/belly demarcation.

Voice Flight call a low, short rolling trill, not particularly musical. Similar to but typically lower and faster than comparable call of Baird's. Breeding-ground display song a low, hollow hooting, increasing in speed as male rockets low over tundra.

Behavior Most feeding in dense marsh vegetation, where birds spread out and difficult to spot. Much like oversized Least Sandpiper, foraging hunched over with head down, walking slowly and both picking and probing.

Habitat Breeds on wet tundra. Migrates and winters primarily on vegetated salt- and freshwater marshes and flooded fields, less commonly on open mudflats.

Range Breeds across Arctic from northern Siberia to west side of Hudson Bay. Winters in southern South America, uncommonly eastern Australia and New Zealand. Migrants most common in midcontinent during both seasons, smaller numbers to both coasts, mostly juveniles. Relatively uncommon west of Rocky Mountains, some migrants from Siberia. Common migrant through Middle America; uncommon in West Indies, mostly in fall.

71.1 Breeding female Pectoral Sandpiper. Medium-small calidridine with short, pale-based bill and yellow legs. Rich brown above with brown, heavily streaked breast. Most brightly colored breeding birds show some reddish on crown, ear coverts, and scapulars. Relatively few feathers black centered (more so in most individuals). Seattle, Washington, USA, May 2003 (Stuart MacKay).

71.2 Breeding male Pectoral Sandpiper. Breeding males with heavily striped breast, showing black feather bases when expanded in display. Cambridge Bay, Northwest Territories, Canada, Jun 2002 (Jim Richards).

71.3 Breeding Pectoral Sandpiper. Heavily worn postbreeding bird, darker above because pale edges mostly worn off; note three freshly molted scapulars. Ventura, California, USA, Oct 2002 (Brian Small).

71.4 Nonbreeding Pectoral Sandpiper. Differs very little from breeding plumage; less likely to show reddish. Note lack of primary projection (clearly present on most birds); perhaps finishing winter wing molt. Bill at pale extreme for species. Bolivia, Mar 1999 (Sam Fried/VIREO).

71.5 Juvenile Pectoral Sandpiper. Very bright plumage, with most of upperparts black centered and rufous fringed; well-defined white mantle and upper scapular lines. Cap can be quite rufous. Ventura, California, USA, Oct 2002 (Brian Small).

71.6 Juvenile Pectoral Sandpiper. Characteristic breast pattern, with lower edge of streaks ending in point. Split supercilium typical of species. Point Reyes, California, USA, Sep 2002 (Peter La Tourrette).

71.7 Pectoral Sandpiper. Typical calidridine flight pattern but quite brown and with inconspicuous wing stripe. Breast pattern distinctive from below. Ventura, California, USA, Sep 2001 (Larry Sansone).

72 Sharp-tailed Sandpiper

(Calidris acuminata)

The limited breeding range of this Siberian representative of Pectoral Sandpiper actually falls within that of its near relative. During the winter each goes to its own hemisphere, overlapping just enough to allow frequent comparisons by birders, when it can be seen that Sharp-tailed is altogether a more showy bird. Although males are slightly larger than females, the Sharp-tailed lacks the dramatic size dimorphism of the Pectoral.

Size Weight 67 g, length 20 cm (8 in), bill 2.6 cm, tarsus 3.0 cm.

Plumages Bill black or dark brown, sometimes with grayish or greenish at base of lower or both mandibles; legs greenish yellow to olive green. *Breeding adult* with rufous, black-streaked crown, fairly conspicuous white supercilium and white eyering, and dark loral stripe expanded into smudge just before eye. Mantle rich brown striped with black, scapulars and tertials broadly fringed with bright rufous; some lower scapulars with whitish outer edges. Face and neck streaked with dark brown; breast buff, heavily spotted with black. Sides and part of belly heavily marked with black chevrons, rest of underparts white. *Nonbreeding adult* much duller, feathers of upperparts dark brown with gray-brown to buff fringes. Some scapulars and/or tertials narrowly edged with rufous in some individuals, as are crown feathers. Breast pale buff with scattered spots and streaks at sides of varying density, in some individuals extending all the way across. *Juvenile* with legs averaging paler, more yellowish, than in adults. Color of upperparts much as in breeding adult, with finely black-streaked, rich rufous cap and wide bright rufous fringes on scapulars and tertials; differs by well-defined white mantle and scapular lines. Broad white supercilium and eyering conspicuous, dark loral stripe conspicuous or not. Underparts more lightly marked than in adult. Breast and foreneck buff, rest of underparts including throat white. Band of fine dark streaks across foreneck, coarser streaks on sides of breast, and fine streaks on sides and undertail coverts.

Identification Superficially much like Pectoral, about same size (less sexual dimorphism in size) and rich brown to gray-brown with short (averages shorter than Pectoral), pale-based bill and pale legs. Bill usually with less pale color on base than Pectoral's and legs often greenish, Pectoral's usually obviously yellow. In all plumages, cap more reddish and white eyering and white supercilium usually more prominent than in Pectoral. *Breeding* adults with upperparts and cap more brightly marked reddish than in Pectoral and underparts quite different, heavily spotted and scalloped with black. Individuals in molt might come close to duplicating Pectoral's breast pattern. *Juveniles* more similar, both with conspicuously striped back, but Sharp-tailed with rich buffy breast, only sparsely streaked and not with strong demarcation between breast and belly. Markings of upperparts usually brighter, but overlap in this, some Pectorals even with rather reddish cap. Juvenile Ruff also liable to be mistaken for juvenile Sharp-tailed, as superficially rather similar. Ruff larger and differently shaped (longer-necked and -legged), upperparts usually look more scaly than striped, and plain face and unstreaked underparts are good distinctions from Sharp-tailed. *Nonbreeding* Pectoral and Sharp-tailed similar enough to present real challenge, both with relatively dull upperparts and streaky breast. Sharp-tailed usually with sparse streaks, but some nonbreeding birds about as heavily streaked as Pectoral; however, markings finer, lines of dots rather than streaks. Fortunately, birds of either species in this plumage virtually nonexistent in North America. If such a bird seen in winter, good description of flight call would be diagnostic.

In Flight Almost identical to Pectoral Sandpiper from above; slightly brighter colors might be evident with good look. Difference in juvenile breast patterns might be evident from below, and birds in spotted breeding plumage should be distinctive.

Voice Flight call a liquid, single *kvik* or *wit* or double *kuvick* or *tuwit*, somewhat similar to call of Barn Swallow. Appears to be less vocal than Pectoral.

Behavior Feeds much like Pectoral by probing and picking in marshes, usually singly or in very small groups in North America (small flocks where more numerous in western Alaska). More likely than Pectoral to forage on mudflats.

Habitat Wet tundra for breeding. Fresh and salt marshes and mudflats for nonbreeding.

Range Breeds in northern Siberia. Winters in Australasia. Migrants regularly to Alaska in fall, less common but still regular south along American Pacific coast to California. Vagrant all across North America, recorded from numerous states and provinces all the way to southern Florida. Few spring records, and relatively few adults in fall; great majority are juveniles.

72.1 Breeding Sharp-tailed Sandpiper. Size and shape of Pectoral Sandpiper but bill darker, legs duller. Upperparts with much rich rufous, including coverts and especially cap. Foreneck and upper breast spotted, lower breast and belly with prominent chevrons. Iwakuni, Yamaguchi Prefecture, Japan, May 1993 (Tokio Sugiyama).

72.2 Breeding Sharp-tailed Sandpiper. Post-breeding individual with much of rufous worn off but still evident in cap. Underparts typical, density of chevrons variable. Ibaraki Prefecture, Japan, Aug 1999 (Norio Kawano).

72.3 Nonbreeding Sharp-tailed Sandpiper. Much duller plumage, with much less reddish above and only markings below fine rows of dots on breast. Cap usually with some rufous, more reddish than in very similar Pectoral Sandpiper. Newcastle, Australia, Mar 2003 (Neil Fifer).

72.4 Nonbreeding Sharp-tailed Sandpiper. Pale buffy breast with sparse rows of dots forming fine streaks characteristic. Reddish cap clearly set off by white supercilium. Newcastle, Australia, Mar 2003 (Neil Fifer).

72.5 Nonbreeding Sharp-tailed Sandpiper. Some individuals have more heavily marked breast, almost as much as Pectoral, and need careful scrutiny. More likely to have fine streaks on posterior sides and undertail. Quite pointed central tail feathers visible (same shape in Pectoral but not other *Calidris*). Newcastle, Australia, Apr 2003 (Neil Fifer).

72.6 Juvenile Sharp-tailed Sandpiper. Much brighter than nonbreeding adult, with broad rufous fringes on upperparts, prominent mantle and scapular lines. Cap bright rufous, supercilium conspicuously white, and breast bright buff, less streaked than in Pectoral. Legs can be as bright as Pectoral's. Fern Hill Wetlands, Oregon, USA, Oct 2002 (Ruth Sullivan).

72.7 Juvenile Sharp-tailed Sandpiper. Older and less bright juvenile, but rufous cap, conspicuous supercilium, and buffy breast still distinctive. Orange County, California, USA, probably Nov 1999 (Brian Small).

72.8 Sharp-tailed Sandpiper. Typical calidridine flight pattern, with rump dark in center and white on either side. Wing stripes inconspicuous as in Pectoral. Clear view of underparts or flight calls necessary to distinguish these two species. Aichi Prefecture, Japan, Oct 2003 (Tomokazu Tanigawa).

73 Purple Sandpiper

(*Calidris maritima*)

This species is common only along the rocky North Atlantic coastline in winter. No better description has been written since Roger Tory Peterson perfectly summed it up in his earlier field guides by "portly, slaty, yellow legs; on rocks." Few see the bird in full breeding plumage, even fewer in the short-lived juvenile plumage.

Size Weight 68 g, length 21 cm (8.25 in), male bill 3.0 cm, female bill 3.5 cm, tarsus 2.3 cm.

Plumages Slight sexual dimorphism in bill length, female's longer. Bill black, grading to yellow base; legs yellow. *Breeding adult* with head, neck, and sides heavily streaked with blackish on mostly whitish ground color, streaks replaced by prominent black spots on breast. Pale supercilium and dusky auricular spot only faintly indicated. Mantle and scapulars with cinnamon-rufous fringes, fading quickly during summer to pale buff or white, producing scalloped effect, then may wear away altogether, producing dull, dark bird by July. *Nonbreeding adult* gray all over except belly, which is white with heavily spotted sides. Feather centers of upperparts and breast slightly darker than edges, producing vague scalloped and striped patterns, respectively. Almost no head pattern but may be limited white markings around or in front of eyes. Vaguely purplish gloss on upperparts that gave bird its name visible only at close range. *Juvenile* much like breeding adult. Dark gray above, crown streaked with rusty to white, mantle and scapulars fringed with rusty, buff, and white, coverts and tertials fringed with white; great variation. Breast gray, spotted and streaked with black; belly white, sides streaked with dark gray to black. Juveniles molt to nonbreeding plumage at high latitude, so this plumage not seen on wintering grounds.

Subspecies Birds breeding on islands of southern Hudson Bay and James Bay distinguished as *C. m. belcheri*, characterized by small size; not distinguishable in field from *C. m. maritima*, which breeds in remainder of North American range. Wintering grounds of the two have not been distinguished.

Identification Short necked, short legged, and rather long tailed, with heavily marked sides in all plumages, most similar to Rock Sandpiper of opposite shore (see that species), but could be mistaken for other small sandpipers. Nonbreeding plumage rather drab like Dunlin but much grayer, less brownish, with yellow legs and shorter, yellow-based bill. Bill shape, gray rather than brown color, and striped sides make for easy distinction from associated Ruddy Turnstones. Breeding adults heavily marked with black on breast and sides and quite reddish above; less reddish than Sanderling or Dunlin, more heavily marked than White-rumped or Baird's, more contrasty than Pectoral. Juveniles primarily on breeding grounds, much shorter legged than Pectoral, darker and yellow-legged in comparison with Baird's and Dunlin, darker and grayer than any of them.

In Flight Dark gray or brown upperparts and breast, conspicuous white wing stripes, dark tail stripe, and gray and white underwings. Darker than others with similar flight pattern, with wing stripes almost as contrasty as much paler Sanderling. Easily picked out among turnstones by less flashy pattern. Probably indistinguishable from darker subspecies of Rock Sandpiper in flight, although wing stripe averages slightly narrower.

Voice Flock calls a single *wit* or *weet-wit* and variations on it, flocks sounding something like Barn Swallows; not very vocal. Breeding-ground display flights accompanied by rapid trills, "moaning" *oo-eeeee* calls, and froglike croaking buzzes, considered very different from Rock Sandpiper songs.

Behavior Almost confined to rocks and cobbles during nonbreeding season, methodically picking and probing for molluscan prey among seaweeds, barnacles, and mussels in small flocks, often with Ruddy Turnstones.

Habitat Moist to dry, often gravelly, tundra for breeding. Rocky shores for nonbreeding. Even more restricted to rocks than Rock Sandpiper.

Range Breeds in eastern Canadian Arctic, only sparingly on mainland; also Greenland and arctic Europe to western Siberia. Winters on Atlantic coasts from Newfoundland to Carolinas, rarely to Florida and Great Lakes; also coasts of western Europe. Migrants coastal. Vagrant elsewhere in interior of continent west to northern Alaska and south to Texas coast.

73.1 Breeding Purple Sandpiper. Stocky calidridine of rocky habitats with short wings typically not reaching tail tip. Overall dark with yellow bill base and yellow legs. Scalloped above and heavily spotted below in this plumage, with indistinct supercilium and usually some reddish on mantle and scapulars. Ocean County, New Jersey, USA, May 1995 (Mike Danzenbaker).

73.2 Nonbreeding Purple Sandpiper. Plain dark gray above and on breast with heavily streaked sides; no supercilium. Dunlin sized but darker. Venice, Florida, USA, Mar 1999 (Brian Small).

73.3 Juvenile Purple Sandpiper. Much like breeding plumage but plainer head, with no supercilium, more neatly scalloped. Rufous present or absent on mantle and scapulars. Most like juvenile Dunlin, which has blackish breast patch, but leg color allows easy distinction. Cleveland, England, Aug 2003 (Wayne Richardson).

73.4 Purple Sandpiper. Much like Dunlin in flight but darker, shorter-billed; look for yellow legs and bill base. Cleveland, England, Oct 1998 (Wayne Richardson).

74 Rock Sandpiper

(*Calidris ptilocnemis*)

Like Purple Sandpiper, this is a northern bird, found commonly on cold rocky coasts but locally on sand and mud substrates. It has decreased in the southern part of its wintering range but is still common in its Alaska stronghold and is one of the most characteristic birds of the region called Beringia. Its very distinct subspecies indicate long isolation into discrete populations.

Size Weight 75 g, length 21 cm (8.25 in), male bill 2.7 cm, female bill 3.1 cm, tarsus 2.3 cm (smaller subspecies).

Plumages Slight size dimorphism, female with longer bill. Bill black, in some individuals yellowish at base; legs olive to gray or black. *Breeding adult* with rufous crown and hindneck; mantle and scapulars black with rufous and whitish fringes and whitish dots. Both brightness and extent of rufous vary considerably (see Subspecies). Tertials dark gray with whitish fringes. Sides of head and throat vary from dark, with poorly defined supercilium, to pale with prominent supercilium and dark lores and ear coverts. Breast mottled dark gray to blackish on white, black more concentrated into solid patch on lower breast. Belly with gray streaks and spots all across it. *Nonbreeding adult* with bill with distinct yellow base, legs dull to fairly bright yellow. Head and breast pale to dark gray, mantle feathers and scapulars black with more or less extensive gray fringes, tertials gray with whitish fringes. More narrowly fringed individuals look black blotched above, more broadly fringed ones entirely gray. Lower breast, sides, and in some individuals, entire belly heavily spotted with dark gray; spots often arranged in lines. Molt into nonbreeding plumage on or near breeding grounds. *Juvenile* with bill and legs as in nonbreeding adult. Crown striped but without rufous. Mantle feathers fringed with buff to rufous. Scapulars and tertials black and coverts dark gray, all fringed with buff. Face and hindneck gray with poorly defined white supercilium. Upper breast gray-brown, lower breast buff, and belly white; breast and sides heavily streaked with brown. Juveniles, like adults, undergo complete body molt on or near breeding grounds.

Subspecies Rather well marked in this species. Most distinct is *C. p. ptilocnemis*, which breeds on Pribilof, Hall, and St. Matthew islands and winters on Alaskan mainland to east. Largest subspecies (averages 10–15 g heavier, bill a few millimeters longer than in other subspecies) and much the palest, easily distinguishable in field from other subspecies that might occur with it in winter and migration. Bright upperpart colors buff rather than rufous in breeding plumage, and breast more lightly and finely marked than in other subspecies in nonbreeding plumage. Smaller and darker *C. p. tschuktschorum* breeds on St. Lawrence and Nunivak islands and from Seward Peninsula to Bristol Bay on mainland. Typically bright in breeding plumage, with broad, rufous fringes on upperparts, pale head and upper breast with well-defined dark ear coverts, and patch on lower breast. Similarly sized and colored in winter, *C. p. couesi* usually more obscurely marked in breeding plumage, with darker head and breast and more obscure ear-covert spot and breast patch. It breeds throughout Aleutians, in Shumagins and on Kodiak Island, and on tip of Alaska Peninsula. Both *tschuktschorum* and *couesi* probably winter south along Pacific coast of Lower 48, as indicated by spring observations and specimens. Readily identifiable *C. p. ptilocnemis* confined to Alaska, with one vagrant to Washington. Finally, Asiatic subspecies *C. p. quarta* of Siberian Commander Islands recorded as vagrant in Alaska. A small subspecies with relatively broad feather edges above in breeding plumage, giving it more uniform upperparts than *tschuktschorum* and *couesi*, but probably could not be confidently distinguished in field.

Identification Short-legged, long-tailed *Calidris*. Usually seen with Black Turnstones and Surfbirds in winter, much paler than former and about shade of latter and, like that species, spotted along sides. Smaller size and typical sandpiper bill distinguish it from both. *Nonbreeding* birds very like Purple Sandpiper, perhaps cannot be confidently distinguished. Although slightly shorter winged and shorter billed than Purple, Rock is also slightly heavier, should look slightly chunkier in field. Gray of breast typically broken into spots at rear, whereas breast of Purple more uniformly gray and averages slightly paler; not easy to see in foraging birds. Rock appears more frequently to have bit of white in front of eye. Pribilof subspecies of Rock Sandpiper has paler, unspotted breast like Purple but is overall paler bird that should not be mistaken for Purple. Bill base and legs of Purple average a bit oranger, those of Rock yellower, but this is average difference, probably not diagnostic. *Breeding*-plumaged birds much like Purple but with breast markings coalesced into black patch on upper belly; Purple merely heavily streaked in that area. Breeding Rock overall much more brightly marked than Purple, with feather edges of upperparts much more buffy or reddish, whereas Purple has these edges narrower and usually whitish. Bill and legs become black in breeding season in Rock, not so in Purple. *Juvenile* much like juvenile Purple but again typically brighter, with wider reddish to buffy feather edges above (Purple with these edges pale buff to white) and distinct buffy wash on breast, lacking in Purple. Flight calls may or may not be different in the two species and should be more thoroughly documented; however, sonograms of these calls look similar. Juvenile Dunlin (with black legs and more distinct belly patch that surrounds legs) also needs to be considered.

In Flight Dark gray or brown upperparts and breast, conspicuous white wing stripes, dark tail stripe, and white underwings. Similar to but distinctly darker than Dunlin. Virtually identical to Purple Sandpiper, but white wing stripe averages wider, even in darker subspecies. Nonbreeding Pribilof subspecies *ptilocnemis* conspicuously paler than other subspecies in flight and has wider white wing stripes, more white under wings, and more extensive white in outer tail feathers, also paler than Surfbird.

Voice Typical flight call *chu-ree*, rolled in middle; also low and conversational single notes and twittering in winter flocks. Mostly silent. Breeding-ground display song a trilled *kurree, kurree, kurree*, then more rapid, buzzier pulsed trills; also cricketlike calls from ground.

Behavior Forages rather like Dunlin with head down, moving slowly forward, but more likely to pick in rock substrates; nevertheless probes into interstices among mussels and barnacles. Associates almost invariably with Black Turnstones and Surfbirds in southern part of winter range, often forms large flocks in northern part. Sand-foraging regular on Pribilofs, where Dunlin uncommon.

Habitat Moist tundra with low shrubs for breeding. Rocky shores for nonbreeding in southern part of winter range, but sand and mud substrates also used in northern part.

Range Breeds in extreme western Alaska and on Bering Sea and Aleutian islands; also scattered on west side of Bering Sea. Winters from southern Alaska to northern California; also from Russia south to northern Japan. Migrants coastal.

74.1 Breeding Rock Sandpiper (*tschuktschorum*). Stocky, short-winged, short-legged, medium-small calidridine. This subspecies brightly marked, with rich rufous crown, mantle, and scapulars (but not coverts and tertials), pale face with dark ear coverts, and black breast patch. Breeds on Alaska mainland. Entirely dark bill and legs during breeding season (not so in Purple Sandpiper). Nome, Alaska, USA, Jun 1998 (Brian Small).

74.2 Breeding Rock Sandpiper (*couesi*). Less brightly marked subspecies with head and breast more obscurely patterned. Breeds on Aleutian Islands. Attu Island, Alaska, USA, May 1984 (Erik Breden).

74.3 Breeding Rock Sandpiper (*couesi*). Small percentage of birds wintering south along Pacific coast seem to be this dull subspecies. Ocean Shores, Washington, USA, May 1980 (Dennis Paulson).

74.4 Breeding Rock Sandpiper (*ptilocnemis*). Brightly marked subspecies like *tschuktschorum* but larger and paler, upperparts light rather than dark rufous. Breeds on Pribilof, Hall, and Saint Matthew islands. Saint Paul Island, Alaska, USA, Jun 1995 (Mike Danzenbaker).

74.5 Breeding Rock Sandpiper (*ptilocnemis*). This subspecies with broad white wing stripe. Wing-up display by breeding male. Saint Paul Island, Alaska, USA, Jun 1990 (Mike Danzenbaker).

74.6 Nonbreeding Rock Sandpiper (*tschuktschorum*). Small, dark gray calidridine of rocky shores with short, yellow-based bill and yellow legs, heavily spotted breast and sides. Shaped like Dunlin, colored like Surfbird. Identical to Purple Sandpiper except breast usually more distinctly spotted. Homer, Alaska, USA, Mar 2000 (Brian Small).

74.7 Immature Rock Sandpiper (*tschuktschorum* or *couesi*). Like nonbreeding adult but heavily fringed coverts, rufous on scapulars of juvenile plumage still present. Ocean Shores, Washington, USA, Oct 1980 (Dennis Paulson).

74.8 Nonbreeding Rock Sandpiper (*ptilocnemis*). Pribilof subspecies much paler than others in this plumage, markings on underparts finer; slightly larger size also evident in direct comparison in mixed wintering flocks on mainland. Homer, Alaska, USA, Mar 2000 (Brian Small).

74.9 Juvenile Rock Sandpiper (*couesi*). Dark like nonbreeding adult but profusely fringed with white above, also much rufous on scapulars. Heavily speckled breast but no solid patch as in breeding adult. Shemya Island, Alaska, USA, Aug 2003 (Martin Renner).

74.10 Juvenile Rock Sandpiper (*ptilocnemis*). As in other plumages, paler than other subspecies, with finer markings below. Yellow bill base and yellow legs distinguish from other calidridines of its size, also fine lines of dots on breast. Note relatively long tail of this species. Saint George Island, Alaska, USA, Aug 1987 (Dennis Paulson).

74.11 Rock Sandpiper (*tschuktschorum* or *couesi*). Like Dunlin in flight but much darker. Typically with Black Turnstones and Surfbirds. Ocean Shores, Washington, USA, Apr 1977 (Dennis Paulson).

74.12 Rock Sandpiper wing specimens. Pribilof-breeding subspecies (lower) with more prominent white wing stripe. Adult with whitish fringes on coverts, juvenile with buff fringes, also rufous fringes on tertials. Adult *tschuktschorum* or *couesi*, Ocean Shores, Washington, USA, Feb 1987; juvenile *ptilocnemis*, Saint Paul Island, Alaska, USA, Aug 1983.

75 Dunlin

(*Calidris alpina*)

One of the most common coastal shorebirds, this species occurs in spectacularly large flocks along the Pacific coast. Able to exploit frigid mudflats and flooded fields with equal success, it winters farther north than many of its relatives. Bright in breeding plumage, with a rich reddish back and black belly patch, it molts dramatically into one of the dullest brown shorebirds in winter.

Size Weight 56 g, length 21 cm (8.25 in), bill 3.8 cm, tarsus 2.6 cm.

Plumages Slight dimorphism in bill length (female's longer) but almost overlap; should be distinguishable in pairs. Bill and legs black. *Breeding adult* with upperparts bright rufous with fine black streaks (when thicker, produce darker effect). Many scapulars and fresh tertials with buff tips that produce slight spangling. Face, neck, and breast may look unmarked white at distance but with fine dark streaks; faintly darker loral stripe. Large, usually solid black, blotch occupies entire center of belly. Autumn molt on or near breeding grounds, so fall migrants arrive farther south in nonbreeding plumage. *Nonbreeding adult* dull, with virtually unmarked gray-brown upperparts, neck, and breast and white belly and undertail coverts. White supercilium extends behind eye, dark loral stripe does not reach eye. By late winter, covert fringes of fresh plumage worn off and breast streaks more distinct. *Juvenile* reminiscent of breeding adult, much brighter than nonbreeding adult. Crown reddish brown, streaked with black; upperparts with rufous or whitish fringes, something like juvenile White-rumped Sandpiper. White mantle lines apparent in some individuals, less often white scapular lines. Whitish supercilium and dark loral stripe. Neck and upper breast buffy brown with dark brown streaks. Belly blotched with dark brown and black corresponding to breeding adult's black belly patch. Like adult, molts into nonbreeding plumage on and near breeding grounds; very rarely, birds with some obvious juvenile plumage show up along migration route.

Subspecies Three subspecies distinguished in North America, several others in Old World. *C. a. hudsonia* breeds across Canadian Arctic and winters in East, migrates on Atlantic coast and Great Plains; *C. a. pacifica* breeds in western Alaska and winters in West, migrates on Pacific coast; and *C. a. arcticola* breeds in northern Alaska and winters in Asia. All basically similar, but *hudsonia* differs in both breeding and nonbreeding plumage by typically having lines of fine spots or streaks along sides, usually lacking in Pacific subspecies. These are visible at reasonably close range, but enough variation that single individuals might be less distinctive than populations. Also tendency for breast streaks of breeding-plumaged *pacifica* to stop short of black belly patch; in *hudsonia* extend right to patch. Overlap probably precludes individual birds from being identified to subspecies, as with so many shorebird subspecies. *C. a. arcticola* slightly smaller than *pacifica*, with slightly shorter bill, and slightly brighter and deeper red above, but probably not distinguishable in field. Both of these subspecies pass through Yukon-Kuskokwim Delta of Alaska in fall migration, one heading south, the other west. A few very small Dunlins seen on Atlantic coast may have been one of European subspecies.

Identification Nothing else like it in breeding plumage. Closest to brightest subspecies (*ptilocnemis* and *tschuktschorum*) of Rock Sandpiper but more uniformly light reddish above (heavily marked with black in Rock) and with discrete black patch on lower belly encompassing legs (upper belly in Rock, barely reaching legs). Colors recall breeding Ruddy Turnstone but very different pattern. Nonbreeding birds among plainest sandpipers ("dun"), among similar-sized species darker than Sanderling, paler and browner than Rock and Purple sandpipers. Larger and shorter-winged than White-rumped and Baird's sandpipers, with less distinct supercilium. Brown breast distinguishes from smaller Western and Semipalmated, longer bill from all of above species. Blackish belly patch of juvenile distinctive on breeding grounds, but somewhat similar to both Purple and Rock juveniles. See Curlew and Broad-billed sandpipers.

In Flight Brown upperparts and breast, white belly, conspicuous white wing stripe, and dark tail stripe. Reddish back and black belly patch distinctive in breeding plumage. Longish, slightly curved bill should be apparent. Of species of similar size and conspicuous wing stripes, darker than Sanderling and paler than Purple and Rock sandpipers.

Voice Flight call a rasping *cheeezp*, fairly loud and far carrying. Not especially vocal. Breeding-ground display song a long, buzzy, pulsing and descending trill, often accompanied by harsh single or double notes.

Behavior Feeds by probing in mud and sand, moving slowly and methodically. Roosts and flies in flocks larger than those of any other shorebird species, when thousands of birds may be in close proximity. At times a flock of hundreds to thousands of birds will move back and forth through sky for an hour or more, changing shape constantly and defying understanding of such energy-consuming behavior.

Habitat Breeding on moist tundra. Nonbreeding on sandy beaches, mudflats, lake shores, and flooded fields.

Range Breeds locally across Low Arctic from western Alaska to west side of Hudson Bay, absent from most Canadian islands. Winters on and very near coasts north commonly to southern British Columbia and New Jersey (small numbers farther north) and south to central Mexico and southern Florida, rarely farther; rare in Bahamas and Greater Antilles, vagrant to northern South America. Migrants mostly coastal but scattered in smaller numbers across interior North America; large flocks in spring in northern Great Plains.

75.1 Breeding Dunlin (*hudsonia*). Medium-small calidridine with long, droopy bill; bill and legs black. Bright in this plumage, with entirely rufous cap, mantle, and scapulars, plain face, big black belly patch. Breeds in central and eastern Canada and winters on Atlantic and Gulf coast. South Padre Island, Texas, USA, May 1998 (Brian Small).

75.2 Breeding Dunlin (*pacifica*). Essentially like eastern subspecies. Both molt out of breeding plumage before late migration south. Breeds in western Alaska and winters on Pacific coast. Nome, Alaska, USA, Jun 1998 (Brian Small).

75.3 Breeding Dunlin (*arcticola*). Shortest-billed subspecies in North America; breeds on north slope of Alaska and winters in Asia. Migrates south before molting. Barrow, Alaska, USA, Jun 1975 (J. P. Myers/VIREO).

75.4 Breeding Dunlin. Japanese migrants arrive early in autumn in breeding plumage. Either *arcticola* from Alaska or *sakhalina* from Russia. Aomori Prefecture, Japan, Jul 1999 (Mike Danzenbaker).

75.5 Nonbreeding Dunlin (*hudsonia*). Plain medium-brown upperparts and brown-streaked breast; long, droopy bill distinctive. Eastern subspecies typically with streaked sides. Ocean County, New Jersey, USA, Jan 1996 (Mike Danzenbaker).

75.6 Nonbreeding Dunlin (*pacifica*). Pacific subspecies typically with no streaks on posterior sides. Penn Cove, Washington, USA, Feb 1993 (Dennis Paulson).

75.7 Juvenile Dunlin (*pacifica* or *arcticola*). Typical juvenile calidridine, with rufous- and white-fringed feathers throughout upperparts. Often mantle and scapular lines evident. Underparts heavily streaked, streaks coalesced almost into patches in same area as adult's belly patch (unique among juveniles). North American individuals molt into first-winter plumage before lengthy southward migration, so not seen by most birders. Peard Bay, Alaska, USA, Sep 1978 (Peter Connors/VIREO).

75.8 Juvenile Dunlin (*hudsonia*). Duller juvenile, unusual in retaining this plumage so far south but beginning molt into nonbreeding plumage. Conneaut, Ohio, USA, Sep 2003 (Robert Royse).

75.9 Breeding Dunlin (*hudsonia*). Long-billed, typical calidridine flight pattern but with striking wing stripe, almost as much so as Sanderling. Rich rufous back distinctive in this plumage. Naples, Florida, USA, May 2000 (Wayne Richardson).

75.10 Nonbreeding Dunlin (*pacifica*). This individual with narrower wing stripe (but not known to vary geographically); very brown coloration with dusky breast distinctive in this plumage. Salton Sea, California, USA, Dec 1989 (Mike Danzenbaker).

76 Curlew Sandpiper

(*Calidris ferruginea*)

A relatively long-legged and long-billed calidridine, this species presents a more elegant appearance than Dunlin, for which it may be mistaken. Its occurrence all across the continent makes that a perennial possibility. However, the beautiful chestnut color of breeding-plumaged adults takes them beyond confusion.

Size Weight 57 g, length 21 cm (8.25 in), bill 3.8 cm, tarsus 3.0 cm.

Plumages Slight plumage dimorphism. Bill and legs black. *Breeding adult* rich rufous overall, contrasting with gray-brown coverts. Crown and mantle broadly streaked with black in male, scapulars with gray-brown to white spots, tertials with white and rufous fringes. Scattered dark bars on sides and undertail coverts. Female duller above with more brown and less rufous and slightly lighter rufous below, with scattered black bars and largely white belly. Supercilium may be distinctly paler than rest of head. Newly molted spring individuals have many underpart feathers tipped with white; these tips worn off in later birds. Upperparts brown to mostly blackish by late summer, underparts entirely dark. *Nonbreeding adult* light gray-brown above, finely streaked; scapulars, tertials, and coverts with narrow white fringes when fresh. Moderately conspicuous white supercilium and brown loral stripe. Cheeks and breast streaked with gray-brown, breast washed with light gray-brown. Uppertail coverts white, with no trace of bars present in breeding birds. *Juvenile* basically like nonbreeding plumage but upperparts distinctly scaly, many feathers with brown subterminal line and whitish fringe. Breast less obviously streaked than in adult but washed with light brown to buff. Belly may also be buffy.

Identification In full breeding plumage could be mistaken for similarly richly colored Red Knot, Red Phalarope, and dowitchers only if slender, slightly curved bill not seen. Also smaller, longer-legged, and longer-necked than knot and dowitchers, not swimming like phalarope. Somewhat like both Dunlin and Stilt Sandpiper in nonbreeding plumage. Distinguished from Stilt by shorter, black legs and slightly more slender bill, from Dunlin by longer legs and neck, more slender and more evenly curved bill (not always obvious without comparison), and paler breast. Curlew also has more distinct supercilium than Dunlin and longer wings projecting beyond tail, with much longer primary projection.

In Flight Medium-small sandpiper with conspicuous white wing stripe, white rump, and gray tail. Differs from White-rumped Sandpiper by more extensive white rump and longer legs projecting beyond tail tip, also distinctly larger. Longish, slightly curved bill good mark. Surprisingly similar to Stilt Sandpiper in flight but legs much shorter, just tips of toes projecting beyond tail (all of toes in Stilt).

Voice Flight call a musical trill, dropping in middle and often written "*chirrip.*" Unlike that of any other regularly occurring North American shorebird, although somewhat like call of Lesser Sand-Plover.

Behavior Forages on mudflats and beaches, also commonly in shallow water, where it probes and stitches with head submerged. In flocks where common, typically single in North America and by itself or loosely associated with other species.

Habitat Wet tundra for breeding. Coastal mudflats, beaches, and lake margins for nonbreeding.

Range Breeds in northern Siberia, rarely east to Alaska. Winters in sub-Saharan Africa, India, Southeast Asia, and Australasia, a surprising longitudinal spread considering limited breeding range. Migrants widespread, east to Bering Sea islands and west rarely to Atlantic coast of North America. Vagrant elsewhere in North America, recorded in many states and provinces, widely in West Indies, and once in Costa Rica.

76.1 Breeding male Curlew Sandpiper. Medium-small calidridine with long, droopy bill, fairly long legs, and long wings. Entirely black bill and legs. Unmistakable in bright breeding plumage, male with chestnut head, neck, and underparts, extensive rufous markings on upperparts. Cyprus, May 2003 (Mike Danzenbaker).

76.2 Breeding female Curlew Sandpiper. Much like male but with more white and black bars below, usually paler overall. Broome, Australia, Apr 2003 (Clive Minton).

76.3 Curlew Sandpiper. Molting from breeding to nonbreeding, some chestnut on belly retained well into fall. Cyprus, Aug 2002 (Mike Danzenbaker).

76.4 Nonbreeding Curlew Sandpiper. Plain plumage, rather Dunlin-like but bill more slender toward tip, legs and wings longer; overall more elegant appearance. Supercilium usually better defined. Probably first-year bird because no breeding plumage this late in spring. Galveston, Texas, USA, Apr 1997 (Brian Small).

76.5 Juvenile Curlew Sandpiper. Like nonbreeding adult but overall darker above, heavily scalloped with pale fringes. Breast usually paler than in adult. Elegant look quite apparent here. Hamko, Finland, Oct 1988 (Henry Lehto).

76.6 Curlew Sandpiper. Conspicuous calidridine wing stripe but differs from most other small species in having white rump (speckled in breeding plumage). Rump patch larger than in somewhat similar White-rumped Sandpiper, wing stripe more conspicuous and legs shorter than in Stilt Sandpiper. With Sharp-tailed Sandpipers. Werribee, Australia, Feb 1997 (Tadao Shimba).

77 Stilt Sandpiper

(Calidris himantopus)

This species is quite distinctive in its barred breeding plumage but not always so obvious in other plumages. Shaped like a yellowlegs and feeding like a dowitcher, it has been fancifully called a hybrid between the two. "Stilt" is a good name for this longest-legged of calidridines, fitted well for its aquatic foraging behavior.

Size Weight 58 g, length 21.5 cm (8.5 in), bill 4.0 cm, tarsus 4.2 cm.

Plumages Bill black, legs dull yellow or yellowish brown in migration to greenish on breeding grounds. *Breeding adult* with reddish crown with narrow white central stripe. Broad supercilium white, lores and ear coverts rufous, with dark smudge before eye. Upperparts black and white striped and blotched, with rufous fringes on some scapulars and tertials. Most of underparts heavily but variably barred with brown. Fringes on upperparts fade to whitish during summer and become narrower, so upperparts more contrasty and scaly in autumn adults. Males average brighter than females but with much overlap, may be distinguishable in some mated pairs. *Nonbreeding adult* with legs dull greenish or brownish yellow to bright yellow. Entire upperparts light gray-brown, tertials darker. Scapulars, coverts, and tertials narrowly fringed with white. Distinct white supercilium, less distinct dark loral stripe. Neck and breast fairly heavily streaked with black. Uppertail coverts less marked than in breeding adults. *Juvenile* with legs as in nonbreeding adult. Superficially similar to nonbreeding adult but upperparts highly patterned, always more so than in adult. Appears either striped or scalloped depending on feather edging. In most highly colored individuals, upperparts largely dark brown, with buff or whitish (occasionally rufous) fringes. White mantle lines often conspicuous, scapular lines not usually evident. Breast more or less streaked with brown, and some individuals finely but heavily spotted and striped on sides of breast, streaks extending down sides. Younger juveniles may be entirely washed with buff below. Uppertail coverts usually white, less often spotted than in adults.

Identification Unmistakable in breeding plumage. Barred below like Wandering Tattler but very differently shaped and otherwise differently colored from that species. Nonbreeding birds dull brownish gray on back and breast, like so many other sandpipers, but long legs, long, droopy bill, and foraging style distinctive. Longer yellowish legs ensure distinction from Curlew Sandpiper and Dunlin, both somewhat similar in bill structure. Juvenile marked like several other calidridines, with streaky or scalloped upperparts, but again combination of droopy bill and long yellowish legs distinguish from White-rumped, Baird's, Pectoral, and Dunlin among common species and Curlew and Ruff among rare species. Can be mistaken for Lesser Yellowlegs, with which it often associates, but look for bill shape and different plumage pattern in all plumages. Stilt often has fine streaks on sides, yellowlegs bars if anything. Wilson's Phalarope paler, with needlelike bill, whiter breast, much shorter legs, and quite different foraging behavior in water.

In Flight Medium-small sandpiper with longish, slightly curved bill, plain back, whitish rump and tail, and long yellowish legs hanging out behind, like a *Calidris* with yellowlegs genes. Very much like Lesser Yellowlegs; bill must be seen or flight calls heard for distinction. Juvenile Stilt often shows faint indication of wing stripe, lacking in yellowlegs. Also like Wilson's Phalarope, which has considerably shorter legs. See Curlew Sandpiper.

Voice Occasionally heard flight call a single low *tew* reminiscent of single Lesser Yellowlegs note but not nearly so loud. Generally very quiet in migration. This call or one like it also may be rolled or trilled, even reminiscent of call of Curlew Sandpiper. Breeding-ground flight display song a buzzy, rhythmically repeated trill, *zurreee zurreee zurrree*, leading into unmistakable braying, *heehaw heehaw heehaw*.

Behavior Forages almost entirely in water, where it probes and stitches in bottom mud much like dowitchers but with rear end more elevated because of longer legs and shorter bill. Where common, may aggregate in large flocks of its own species but often forages with both yellowlegs and dowitchers. Uses both fresh- and saltwater wetlands where common but mostly avoids marine habitats on west coast.

Habitat Breeders on moist tundra. Nonbreeders on salt lagoons, freshwater ponds and lakes, and flooded fields.

Range Breeds across Low Arctic from northern Alaska to west side of Hudson Bay. Winters from central Mexico to southern South America, but outliers north to southeast California, south Texas, southern Florida, and Caribbean islands. Migrants mostly midcontinent in spring, east to Atlantic coast in fall, least common west of Great Plains where mostly juveniles present.

77.1 Breeding Stilt Sandpiper. Long-legged, medium-small calidridine with fairly long, droopy bill. Bill black, legs yellow to greenish. Rufous cap and cheeks, along with heavily barred underparts, quite distinctive in this plumage. McAllen, Texas, USA, May 1998 (Brian Small).

77.2 Nonbreeding Stilt Sandpiper. Plain gray-brown above and on breast, white below. Droopy bill and long, pale legs distinctive. Salton Sea, California, USA, Jan 2000 (Rick and Nora Bowers/VIREO).

77.3 Juvenile Stilt Sandpiper. Juvenile with rufous, buff, and white fringes above, may or may not have mantle and scapular lines (some look more striped, some more scalloped). Long yellow legs distinguish from any other calidridine except much chunkier, shorter-billed Ruff. Damon Point, Washington, USA, Aug 1995 (Ruth Sullivan).

77.4 Juvenile Stilt Sandpiper. Brightly plumaged individual with buffy breast. Dowitcher-like feeding behavior very good behavioral clue for this species. Note long wings and primary projection. Conneaut, Ohio, USA, Aug 2002 (Robert Royse).

77.5 Juvenile Stilt Sandpiper. Some juveniles considerably duller, with little rufous or buff. Warren County, Kentucky, USA, Sep 2003 (David Roemer).

77.6 Immature Stilt Sandpiper. Beginning molt into first-winter plumage, approaching dullness of nonbreeding adult but still with conspicuous fringes. Cape Race, Newfoundland, Canada, Sep 1991 (Bruce Mactavish).

77.7 Stilt Sandpiper. Calidridine with plain wings and whitish rump and tail. May look very much like Lesser Yellowlegs in flight, differs by droopy bill, shorter (no tarsus visible) and usually duller legs. Jamaica Bay, New York, USA, Aug 2000 (Michael Stubblefield/VIREO).

78 Spoon-billed Sandpiper

(*Eurynorhynchus pygmeus*)

This little sandpiper, formerly called Spoonbill Sandpiper, is basically a stint more unusually adapted for feeding than most members of its family. With its modified bill, it is not surprising that it feeds like its much larger namesakes. Its declining population, already small, is cause for great concern.

Size Weight 26 g, length 15 cm (6 in), bill 2.1 cm, tarsus 2.0 cm.

Plumages Bill and legs black. *Breeding adult* quite like Red-necked Stint in coloration, with rich orange-brown head and neck, dark spots on sides of breast, and extensive reddish fringes on feathers of upperparts. *Nonbreeding adult* pale gray-brown above as in black-legged stints; supercilium split, very broad in front of eye. *Juvenile* upperparts spangled; mantle, scapulars, tertials, and coverts with extensive black bordered by brown, fringes whitish. Scapulars may have reddish fringes, often whitish mantle and/or scapular stripes evident. Supercilium as in nonbreeding adult, breast also but sometimes buff tinged.

Identification Fairly typical stint, with black legs and moderately long wings, but outrageous bill, rather long and heavy and abruptly widened into spoon-shaped tip, has placed it in different genus. From side, "spoon" may not be obvious, but bill looks heavier than those of other stints because of it. Beware, however, of other stints with sand or mud on bills. Breeding adult much like Red-necked Stint, most prominent difference entirely rufous head (Red-necked usually has white supercilium) and averages more spots behind breast. Nonbreeding adult and juvenile colored much like equivalent plumages of Red-necked Stint (and thus Semipalmated Sandpiper) but have strongly marked split supercilium and extensively white forehead, more white than in any stint. Extensive white isolates dark loral spot before eye (spot virtually absent in some) and brown ear coverts as distinct dark cheek patch. Some juveniles starkly marked like juvenile Sanderling, others with rufous on scapulars more like juvenile Western Sandpiper.

In Flight Entirely stintlike, but heavy bill should be obvious with good look.

Voice Flight call reported as "a shrill *wheet*" and "a quiet rolled *preep*."

Behavior Feeds by sweeping bill back and forth like little spoonbill or picking prey from shallow water or wet mud; not surprisingly, no probing. Perhaps because so uncommon, usually seen singly, occasionally in small flocks mixed in with other small species.

Habitat Coastal tundra for breeding, coastal mudflats for nonbreeding.

Range Breeds in northeastern Siberia. Winters in Southeast Asia. Migrants mostly coastal. Vagrant to North America, few spring and fall records from western and northern Alaska, two spring records from Alberta, and one fall record from British Columbia.

78.1 Breeding Spoon-billed Sandpiper. Small calidridine, unmistakable if spatulate bill clearly seen. Very like Red-necked Stint in this plumage; rufous on upper-parts and brown spots behind rufous upper breast often more extensive, making even more striking bird. Wings long as in Red-necked Stint. Beringovski, Russia, Jul 2000 (Christopher Schenk).

78.2 Nonbreeding Spoon-billed Sandpiper. Plain gray-brown above, white below, again like Red-necked Stint. Other than easy distinction by bill shape, also has more extensive white on head than any stint, with very white forehead, wide split supercilium. Kagoshima Prefecture, Japan, Dec 1995 (Tomokazu Tanigawa).

78.3 Juvenile Spoon-billed Sandpiper. This plumage can be brightly marked with rufous above, more so than Red-necked Stint and almost as vivid as Little Stint. Extensive white on head as in nonbreeding. Aomori Prefecture, Japan, Sep 1999 (Mike Danzenbaker).

78.4 Juvenile Spoon-billed Sandpiper. Some juveniles show no rufous at all. Wing stripe as in black-legged stints. Mie Prefecture, Japan, Oct 2000 (Norio Kawano).

78.5 Juvenile Spoon-billed Sandpiper. Just like black-legged stints in flight, with similar wing stripe. Bill must be seen for positive identification; spoon not always evident from side, but bill longer than in any stint but Western Sandpiper. Ishikawa Prefecture, Japan, Sep 2003 (Tomokazu Tanigawa).

79 Broad-billed Sandpiper

(*Limicola falcinellus*)

Like Spoon-billed Sandpiper, only because of its odd bill is this species not a *Calidris*. Whereas the Spoon-billed's bill seems specialized for a particular diet or foraging method, the Broad-billed employs all the feeding methods of its calidridine relatives. Its somewhat snipelike plumage seems anomalous in a bird that feeds on mudflats.

Size Weight 37 g, length 17 cm (6.5 in), bill 3.1 cm, tarsus 2.2 cm.

Plumages Bill black, legs olive to black. *Breeding adult* richly colored, like most vividly marked Least Sandpipers, with buff-fringed black feathers of upperparts and light brown breast finely streaked with black, including arrowheads on sides. Pale supercilium dramatically split into two stripes. *Nonbreeding adult* similar but light gray-brown above, scapulars and tertials with black centers to make patterned upperparts. Breast lightly to heavily streaked. *Juvenile* legs paler, may be yellowish. Strongly striped above with whitish mantle and scapular stripes, feather margins varying to buff or rufous in some individuals, breast also lightly to heavily streaked. Split supercilium also conspicuous in this plumage.

Subspecies Vagrants to Pacific North America presumably Siberian-breeding *L. f. sibirica*, which in comparison with presumed Atlantic-coast vagrant Scandinavian-breeding *L. f. falcinellus* has more conspicuous rufous fringes on upperparts and cinnamon wash across breast in both breeding and juvenile plumages (but some Siberian birds dull). Siberian subspecies said to have less conspicuous upper branch to split supercilium, but this not borne out in photos.

Identification Often compared with Dunlin, and not much smaller than smallest European ones, but conspicuously smaller than American Dunlins and looks like large, heavy-billed, black-legged stint as much as anything. Slightly smaller than Baird's and White-rumped sandpipers, bill both longer and thicker and wings considerably shorter. Large bill wider toward tip than in other small sandpipers, with slight droop or even kink at end, somewhat larger than that of Western Sandpiper and reminiscent of Dunlin. Each plumage quite different from comparable Dunlin plumage. Looks striped in all plumages, as fringes run all along outer and inner edges of pointed feathers of upperparts, breast usually distinctly streaked. Distinguished by distinctively striped head, broad pale stripe on either side of crown not present on any other small sandpiper. This is "split supercilium" of some other sandpipers taken to extreme, giving it almost snipe-striped head. Also, in many birds loral stripe expands behind eye, forming strong, dark cheek stripe that is distinctive. In breeding plumage, has been compared with Jack Snipe, but back stripes much less vivid, usually feeds more in open. Split supercilium less evident in nonbreeding plumage, but stint look dispelled by large bill, coarsely streaked upperparts, and finely streaked breast.

In Flight Looks like large stint with heavy bill, wing stripe not quite as conspicuous as in Dunlin.

Voice Flight call a dry, buzzing trill, *brrzzeeet*, high pitched and ascending; also single *tzit* calls.

Behavior Forages by probing like Dunlin, moving rather slowly along mudflats, along shores, or in shallow water with head submerged; perhaps bill is adapted for soft mud. However, also picks prey in many situations. Usually in small numbers mixed with other sandpipers, even where common. Quite a skulker for its group, almost snipelike.

Habitat Tundra for breeding. Mudflats, meadows, and pond margins during rest of year. Definitely prefers mud substrate for foraging.

Range Breeds in Scandinavia and scattered across northern Siberia. Winters from Arabian Peninsula east to northern Australia. Migrants anywhere in between. Vagrant to North America; few fall records from Aleutians and one from New York.

79.1 Breeding Broad-billed Sandpiper (*falcinellus*). Small calidridine with long, heavy bill and prominently striped head. Legs olive. In this plumage, dark upperparts with rufous and buff fringes, prominent mantle and scapular lines (snipelike even in this), heavily streaked breast, and dark chevrons on sides. European-breeding subspecies. Salo, Finland, May 1993 (Henry Lehto).

79.2 Breeding Broad-billed Sandpiper (*falcinellus*). Thick bill and prominent stripes on head and upperparts apparent, also especially dark facial stripe. Rufous edgings on upperparts wear off during summer, leaving very dark bird. Cyprus, Sep 2000 (Mike Danzenbaker).

79.3 Breeding Broad-billed Sandpiper (*sibirica*). Siberian subspecies paler overall, more rufous above, breast stripes less bold. This dull individual perhaps not in full plumage. Slightly kinked appearance of bill characteristic. Port Hedland, Australia, Apr 1987 (Clive Minton).

79.4 Nonbreeding Broad-billed Sandpiper (*sibirica*). Upperparts plain gray-brown with some black feather centers, breast sparsely streaked. Conspicuous head stripes and bill shape distinctive. Tushima, Nagasaki Prefecture, Japan, Apr 1980 (Tokio Sugiyama).

79.5 Nonbreeding Broad-billed Sandpiper (*sibirica*). Broad bill and split supercilium evident. In flight with typical calidridine pattern, like large stint, but heavy bill should be apparent. Distinctly smaller than American Dunlins. Okinawa Prefecture, Japan, Sep 2001 (Tomokazu Tanigawa).

79.6 Juvenile Broad-billed Sandpiper (*sibirica*). Bright rufous-fringed upperparts and conspicuous white mantle and scapular lines as in many other juvenile calidridines. Coverts brightly fringed, breast more finely streaked than in breeding. Split supercilium always conspicuous. Aomori Prefecture, Japan, Sep 1999 (Mike Danzenbaker).

79.7 Juvenile Broad-billed Sandpiper (*sibirica*). Much duller individual but with essentially same field marks. Nishio, Aichi Prefecture, Japan, Sep 1987 (Tokio Sugiyama).

80 Buff-breasted Sandpiper

(Tryngites subruficollis)

This is another of the sandpipers committed to an upland way of life, whether on tundra, pampas, or prairie. Its short bill, entirely buff underparts, and bright yellow legs give it distinction. Like the Eskimo Curlew with which it shared a migration route, its populations have been reduced by widespread conversion of natural grasslands to agriculture.

Size Male weight 69 g, female weight 56 g, length 19 cm (7.5 in), bill 2.0 cm, tarsus 3.1 cm.

Plumages Slight dimorphism in size, male larger. Bill black, legs bright yellow to orange-yellow. *Adult* evenly and coarsely patterned above; all feathers with narrow to broad light brown fringes, some scapulars with white tips. Underparts extensively buff. Heavily worn fall adults with narrower and more faded fringes. *Juvenile* with feathers of upperparts narrowly fringed with pale buff; darker, more narrowly fringed than adults, every feather clearly outlined and giving scalloped effect. Underparts as in adult but averaging slightly paler, especially on belly.

Identification Most similar to *Calidris* sandpipers such as Baird's, which has finely streaked buffy breast and white belly rather than entirely buff underparts. Buff-breasted only sandpiper in size range with no conspicuous markings on sides of head, dark eye isolated in field of buff. Other species have pale supercilium and dark lores, and most have black or duller yellow legs, not bright yellow of Buff-breasted. Perhaps closest is juvenile Ruff, with buffy breast and pale lores, but that is considerably larger and white-bellied and has dark streak behind eye.

In Flight Smallish sandpiper with short bill, plain brown upperparts, faintly indicated tail pattern of darker subterminal bar, and whitish tips to most feathers. Entirely buff below, with conspicuously contrasting white underwings. Distinctly smaller in flight than species such as Upland Sandpiper and golden-plovers with which it shares upland habitat. More likely to be mistaken for small sandpipers such as Baird's and Pectoral, but no trace of white on sides of rump and tail, and no hint of white wing stripe. Underwing beautifully patterned for wing-up displays on breeding grounds.

Voice Usual flight call a low, single *tu* or short trill something like that of Baird's Sandpiper or turnstone chatter. Generally silent; calls soft enough to be scarcely audible at moderate distances. Also largely mute on breeding grounds (males make soft ticking sound during displays); as in Ruff, visual display adequate for advertisement.

Behavior Typically upland feeder, walking rapidly through open grassland, plowed fields, or upper beaches with head bobbing like pigeon. Feeds by picking prey from surface and may feed ploverlike by watching for prey, then running to it. Rarely moves out onto mudflats. Forms small flocks or occasionally flies with similar-sized sandpipers.

Habitat Breeders on dry tundra. Nonbreeders mostly on grassland, wandering to beaches, mudflats, golf courses, and rice and other agricultural fields.

Range Breeds primarily in Canadian Arctic Archipelago, also northern Alaska and Northwest Territory. Winters in southern South America. Migrants primarily in mid-continent spring and fall, some south across Atlantic in fall. Also south through Middle America (mostly fall in Central America), juveniles more widespread but more common in Atlantic than Pacific states. Rare in Lesser Antilles in fall, vagrant elsewhere in West Indies.

80.1 Adult female Buff-breasted Sandpiper. Medium-small calidridine with bright yellow legs. Dark above with pale fringes on most feathers, entirely buffy below with plain face. Cambridge Bay, Northwest Territories, Canada, Jun 1975 (Dennis Paulson).

80.2 Juvenile Buff-breasted Sandpiper. Like adult but many feathers of upperparts extensively pale, posterior underparts paler. Ventura, California, USA, Sep 1995 (Brian Small).

80.3 Juvenile Buff-breasted Sandpiper. Flight pattern plain wings and tail above, obviously buffy below. The Netherlands, Sep 2000 (Marten van Dijl).

81 Ruff

(*Philomachus pugnax*)

Although not seen in North America, the spectacular communal lek displays of the Ruff make it especially noteworthy. The male polymorphism is associated with a promiscuous mating system, the pale birds "satellites" that associate with darker birds and at times are able to mate with females attracted by the display. The females are often called "reeves."

Size Male weight 180 g, length 29 cm (11.5 in), bill 3.5 cm, tarsus 5.0 cm; female weight 109 g, length 24 cm (9.5 in), bill 3.1 cm, tarsus 4.3 cm.

Plumages Great dimorphism in size (male dramatically larger) and plumage. Bill in male usually pinkish or orange but in some individuals gray or yellow, often with black tip; bill in female black, in some with reddish base. Legs usually orange or pink. In addition to developing complex plumage, breeding males lose feathers on forehead and lores and develop "warts" there, usually yellowish or reddish. *Breeding male* with individual variation that almost defies description. Head (with tufts), neck ruff, breast, and back all vary in color, with ground color black, rufous, buff, white, or some lighter or darker variation on these themes. Ground color may be overlaid with barring or spotting of almost infinite variability in shape and size; markings vary from coarse, producing contrasty pattern, to very fine, producing some apparently intermediate color. Plumage areas vary independently, so many combinations possible. Scapulars and tertials typically rufous or gray-brown, extensively tipped or more finely barred with dark brown or black; often finely vermiculated along much of length, varying more rarely to black or black and white. Lower belly always white, but breast color and pattern may extend down sides to base of legs. Even late in migration (late April), many males still without full display plumage. *Breeding female* also more variable than other shorebird species, although much less so than breeding males. Typically upperparts vary from cinnamon or rufous to gray. Mantle feathers, scapulars, and tertials variably marked with black bars or wider subterminal markings. Neck and breast similarly vary from rufous to buff to gray, relatively plain or fairly heavily spotted or barred with brown or black. A few individuals approach males in intensity of coloration, with largely rufous or blackish foreparts or vividly striped upperparts, but none shows head tufts or neck ruffs or bare skin. *Nonbreeding adult* with bill black, in some individuals brown at base. Sexes look similar in this plumage. Entire upperparts plain gray-brown with darker feather centers. Some individuals with fairly wide, prominent black bars on some or all mantle feathers, scapulars, coverts, and tertials. Neck and upper breast light gray-brown, faintly to moderately barred with darker gray-brown. Some males may be extensively white on neck and breast, rarely with entirely white head and breast, perhaps early acquisition of part of breeding plumage. *Juvenile* with bill as in nonbreeding adult, legs generally duller than in adults, greenish to yellowish. Substantial individual variation confuses issue of leg color, and legs apparently change color with age in some individuals but not others. Overall impression can be gray or buffy brown, with buff or rufous fringes giving striped effect above. Bright markings of upperparts allow ready distinction from nonbreeding adult. Poorly defined pale buff or whitish supercilium, narrow dark postocular stripe. Neck streaked and breast in some individuals obscurely barred with slightly darker gray-brown.

Identification Ruff sexes in very different size categories, male about bulk of Greater Yellowlegs, female of dowitcher size. Unmistakable in male *breeding* plumage (but so striking it may startle someone who has never seen one before) but fits in general category of large calidridine sandpiper in all other plumages. Proportions different than in similar-sized species, looking plump perhaps because of short bill and seemingly small head. Neck fairly long, legs somewhat long for calidridine, so very differently shaped from knot species, which are also grayer, not as brown, in *nonbreeding* and *juvenile* plumages. Richly colored juvenile most like Pectoral and Sharp-tailed sandpipers but distinctly larger and with scaly rather than striped back. Differs further in breast buffy to gray but completely without streaks (streaks prominent in Pectoral, scattered in Sharp-tailed). Only other calidridine lacking dark loral stripe is Buff-breasted, which also lacks postocular stripe; thus stripe behind eye but no stripe in front of eye diagnostic of Ruff. Richly colored juveniles somewhat like golden-plovers in same plumage but readily distinguished by sandpiper bill, orange legs, and different foraging behavior.

In Flight Medium sandpiper, brown above and white below with typical calidridine flight pattern much enhanced, white as two conspicuous ovals on either side of dark rump stripe. Inconspicuous white wing stripes. Breeding male much more garishly colored, different morphs mostly white, mostly rufous, mostly black, or combinations thereof. Most like Pectoral of common sandpipers (and Sharp-tailed of rare sandpipers) but white rump ovals distinctive, and toes project well beyond tail tip. Flight style characteristic, with slower and deeper wing beats than similar-sized species, giving appearance of more casual approach to flying.

Voice Low whistled or grunting calls given occasionally, but one of quietest shorebirds.

Behavior Very active forager in almost any open marine, freshwater, or terrestrial habitat, walking steadily or running rapidly and foraging visually by picking prey from land or shallow water. Characteristic hunched posture, sometimes back feathers fluffed up, when foraging slowly. Forms large flocks where common, usually seen singly in North America. Food habits varied; subsists entirely on rice in some wintering areas.

Habitat Wet tundra for breeding, usually near water, also wet meadows and marshes farther south. Nonbreeders on coastal mudflats and beaches, lake margins, flooded fields, and marshes.

Range Breeds across Eurasia from western Europe to far eastern Siberia at high and moderate latitudes; bred once in Alaska. Winters in sub-Saharan Africa and India, less commonly farther east. Migrants over wide area, surprisingly frequent on both coasts and across interior of North America, whence recorded from many states and provinces and throughout West Indies; regular on Barbados. Very rare anywhere on neotropical mainland, records from Pacific coast in Baja California, Guatemala, and Costa Rica, both coasts of Panama; few to South America. Interestingly, more common in spring on Atlantic coast, in fall on Pacific coast, thus different plumages seen there; has wintered on both coasts.

81.1 Breeding male Ruff. Largish calidridine with plump appearance, short bill, and fairly long legs. Breeding males as showy as any grouse in communal display, spectacularly endowed with paired head tufts and wide neck ruff, bare face. Scapulars and tertials broadly barred. This individual at dark end of variation. Siberia, Jun 1994 (Clive Minton).

81.2 Breeding male Ruff. Black and rufous combinations common. Chiba Prefecture, Japan, Apr 2001 (Norio Kawano).

81.3 Breeding male Ruff. Rufous may vary from minor plumage component to entire upperparts and breast, finely or coarsely barred or unbarred. Mietoinen, Finland, May 1992 (Henry Lehto).

81.4 Breeding male Ruff. White may also be important component of breeding colors; light males are called "satellites," subordinate to dark males at display sites. Varanger Peninsula, Norway, Jun 1981 (Markku Huhta-Koivisto).

81.5 Breeding male Ruff. May be strikingly colored early in season, before development of head tufts, ruff, and bare face. Ibaraki Prefecture, Japan, Apr 2001 (Norio Kawano).

81.6 Breeding female Ruff. Typical female with patterned upperparts and scattered black spots on brownish underparts. Legs usually orange during breeding. Female often called "reeve." Paimio, Finland, May 1990 (Henry Lehto).

81.7 Nonbreeding Ruff. Female and two males. Plain brown above in this plumage, with barred tertials. Obscure streaks on sides. Sexual dimorphism in size greater than in any other shorebird; males with bulk of Greater Yellowlegs, females of dowitcher. Norfolk, England, Feb 1998 (Colin Bradshaw).

81.8 Nonbreeding Ruff. Many nonbreeding males have white on the head or even head, neck, and entire underparts. Species often distinguished by actively running foraging style. Diksmuide, Belgium, Jun (Verbanck Koen).

81.9 Juvenile Ruff. Mantle feathers often raised while feeding, more likely in this species than other shorebirds. Larger and longer-legged than species such as Sharp-tailed and Buff-breasted sandpipers with similarly buffy breast. Rotterdam, The Netherlands, Aug 2003 (Norman van Swelm).

81.10 Juvenile Ruff. Upperparts dark, conspicuously fringed with buff on all feather tracts, usually producing scaly effect. Foreneck and breast gray-brown to rich buff. Legs greenish or yellow. Fairly short bill, small-headed look, plain face, and longish legs (reminiscent of Upland Sandpiper) all distinguishing features. Kaarina, Finland, Aug 1996 (Henry Lehto).

81.11 Adult Ruff. Calidridine flight pattern with narrow but obvious wing stripe, extralarge white outer uppertail coverts producing large white ovals on either side of rump and tail; sometimes overlap to produce white across tail. Note sexual size dimorphism; four males and two females. Also variation in bill and foot color, possibly related to age. Lake Koka, Ethiopia, Feb 1997 (Göran Ekström).

Dowitchers (Limnodromini)

Dowitchers are long-billed, short-legged sandpipers, the two American species (of three worldwide) similar in appearance and behavior. Foraging is sewing-machinelike, the bill inserted and withdrawn rapidly from the substrate as the bird moves slowly forward. Heads down much of the time, dowitchers are probably well served by the more alert plovers and tringines with which they associate. Something that looks intermediate between a Bar-tailed Godwit and a dowitcher might be an Asian Dowitcher (*Limnodromus semipalmatus*), a possible vagrant to Pacific North America.

82 Short-billed Dowitcher

(Limnodromus griseus)

Dowitchers are distinctive because of their tight flocks, long bills, and methodical feeding probes. Heavily mottled buff and black above and largely reddish to orange below in breeding plumage, they molt in autumn to dull gray-brown suitable for winter mudflats.

Size Weight 105 g, length 27 cm (10.5 in), male bill 5.7 cm, female bill 6.2 cm, tarsus 3.7 cm.

Plumages Bill black, may be tinged gray or greenish at base; legs light green. *Breeding adult* with upperparts mottled and barred black and rufous, with buffy to whitish fringes that form conspicuous mantle lines and indistinct scapular lines. Loral and short postocular stripe dark, fairly prominent supercilium rufous or partially white. Underparts entirely rufous or with extensive white posteriorly. Sides and undertail coverts spotted and barred; breast in most individuals spotted, bars or spots may extend over entire underparts in more heavily marked individuals, whereas most lightly marked individuals show scarcely any markings. *Nonbreeding adult* with bill greenish gray at base. Upperparts and breast brownish gray, breast with scattered dark spots; belly white, sides with broad gray bars. Distinct white supercilium, dark loral stripe, and short, dark postocular stripe. Posterior sides and undertail coverts spotted and barred with black. *Juvenile* with bill more extensively pale greenish gray at base than in adult. Plumage brighter than in nonbreeding adult, with darker crown and dark, buff-striped mantle. Blackish scapulars and tertials with reddish fringes and submarginal markings. Breast gray to rich buff, buff in some individuals extending onto belly. May be rich rufous below on breeding grounds, this color fading quickly during migration; early-migrating juveniles might be mistaken for breeding adults because of this. Usual contrast between buff breast and white belly and cleanly striped appearance of upperparts allows easy distinction of juveniles from breeding adults.

Subspecies Probably long separated into three breeding populations, has differentiated into three named subspecies in North America. *L. g. griseus* breeds in Canada east of Hudson Bay and migrates down Atlantic coast to winter primarily in Caribbean. *L. g. hendersoni* breeds in Canada west of Hudson Bay and migrates throughout eastern North America, wintering along Atlantic and Gulf coasts and farther south. *L. g. caurinus* breeds in southern Alaska, migrates mostly along Pacific coast, and winters from California south along that coast. First two and second two subspecies pairs known to overlap in migration, but no definite records of *griseus* on Pacific coast or *caurinus* on Atlantic coast. In breeding plumage typical individuals of each may be recognizable in field. *L. g. griseus* tends to have heavily marked breast and mostly white belly; also marginally smallest and shortest-billed subspecies. *L. g. hendersoni* tends to have wide, pale fringes on feathers of upperparts and evenly colored reddish underparts with scattered small dark spots and fairly extensive bars. *L. g. caurinus* tends to have narrower, darker fringes on upperparts, dark spots of underparts concentrated on breast, and some white on belly. Juveniles of *hendersoni* and *caurinus* differ in ways similar to adults, former having brighter, wider markings on upperparts and averaging paler below. In *griseus,* dark tail bars tend to be wider than white, in *hendersoni* and *caurinus* dark bars equal to white or narrower. It must be added that each subspecies is variable, and identification of single birds out of their normal range is problematic.

Identification In breeding and nonbreeding plumages superficially similar to knots but much longer billed. When bill hidden in sleeping or probing birds, distinguished from Red Knot in breeding plumage by browner back, lack of contrast between white undertail and rich red of rest of underparts. Distinguished from nonbreeding and juvenile knots under same conditions by overall darker coloration and distinctly darker sides. Much smaller and shorter-legged than similar-feeding godwits. See Long-billed Dowitcher for differences between these two very similar species. Many Short-billeds in spring are in nonbreeding or mixed plumages, probably all immatures, whereas this seems to be rare in Long-billed, which may mature more quickly.

In Flight Medium-small sandpiper with long bill, brown above with white rump patch extending up lower back as point; tail not as pale as in most white-rumped species. Whitish line on rear edge of wings. Reddish below in breeding plumage, other plumages with gray to buffy breast, white belly.

Voice Flight call *tututu,* staccato series of low, musical notes a bit faster than but similar to those of Lesser Yellowlegs. Series of three notes typical, with some variation to two or four. Some birds give notes in longer, quicker series. Quite distinct from call of Long-billed and best "field mark" to distinguish them. Statements in literature that individuals occasionally give call of other species probably erroneous. Breeding-ground flight display song similar in both species but with minor differences, a broken trill ending in buzzy *dowitcher*. First part of song in Long-billed characterized by series of doubled notes. Dowitchers commonly sing in flight in spring.

Behavior Prototypical stitching sandpiper, long bill worked up and down in substrate like sewing-machine needle as bird moves slowly across mudflats and interior marshes. Slow forward movement distinctive, in contrast with many other sandpipers. Forms large roosting flocks but much smaller flocks in flight. Commonly flocks with Black-bellied Plovers, Willets, Red Knots, and Dunlins.

Habitat Breeders in muskegs in boreal forest zone. Nonbreeders on mudflats and borders of freshwater lakes; when on sandy beaches usually at roost.

Range Breeds in three taiga populations: southern Alaska, north of Canadian prairies from northeastern British Columbia to Hudson Bay, and in northern Quebec and Labrador. Winters along southern coasts from California and North Carolina south through Middle America and West Indies to Ecuador and northern Brazil. Migrants widespread across continent but concentrated on Pacific and Atlantic coasts and Great Plains, corresponding to three subspecies (which see for details).

82.1 Breeding Short-billed Dowitcher (*griseus*). Long-billed, medium-small sandpiper with reddish underparts, heavily spotted and barred sides in this plumage. Atlantic coast subspecies with least amount of rufous below, mostly white lower breast and belly. Cape May, New Jersey, USA, May 1993 (Bruce Mactavish).

82.2 Breeding Short-billed Dowitcher (*griseus*). Late in spring, but minimal amount of rufous below indicates this subspecies wandering inland. Very prominent bars on sides of breast as in Long-billed Dowitcher, but black and white bars elsewhere characteristic of Short-billed. Warren County, Kentucky, USA, Jun 2002 (David Roemer).

82.3 Breeding Short-billed Dowitcher (*griseus*). Irregular barring on tail characteristic of species, very unlikely in Long-billed. Warren County, Kentucky, USA, Jun 2002 (David Roemer).

82.4 Breeding Short-billed Dowitcher (*griseus*). This tail pattern could be seen in either species, as eastern Short-billed can have wide black bars just like Long-billed. Central feathers with no rufous, however, indicate Short-billed. Warren County, Kentucky, USA, Jun 2002 (David Roemer).

82.5 Breeding Short-billed Dowitcher (*hendersoni*). Midcontinent subspecies usually with underparts rufous, markings restricted to scattered black dots as well as side bars. Note numerous coverts and tertials retained from nonbreeding. Upper Coast, Texas, USA, May 1999 (Brian Small).

82.6 Breeding Short-billed Dowitcher (*hendersoni*). More thoroughly molted individual on breeding grounds. Many sandpipers of boreal forest perch in trees. Churchill, Manitoba, Canada, Jul 1999 (Anthony Mercieca).

82.7 Breeding Short-billed Dowitcher (*caurinus*). Pacific coast subspecies somewhat intermediate, usually less white on belly and fewer ventral markings than *griseus* but more white and more ventral markings than *hendersoni*. Spots rather than bars on sides of breast good quick way to distinguish from Long-billed. This individual rather like *hendersoni*, not surely known to occur on Pacific coast. Ventura, California, USA, May 2000 (Brian Small).

82.8 Breeding Short-billed Dowitcher (*caurinus*). Typical *caurinus*, also showing that some Short-billeds can have bars on sides of breast. West-coast birds typically have distinct primary projection, lacking in Long-billed. Ocosta, Washington, USA, May 2003 (Stuart MacKay).

82.9 Breeding Short-billed Dowitcher (*caurinus*). Post-breeding individual beginning autumn molt, usually earlier than Long-billed. Huntington Beach, California, USA, Aug 1999 (Joe Fuhrman).

82.10 Nonbreeding Short-billed Dowitcher. Plain gray-brown above and on breast, with barred sides. Distinguished from Long-billed by plain mantle and scapulars, irregularity and usually higher contrast of side barring. Also should have more speckling on breast. Could be *griseus* or *hendersoni*. Lee County, Florida, USA, Mar (Clair Postmus).

82.11 Nonbreeding Short-billed Dowitcher (*caurinus*). Typical field marks shown, including side bars better defined than in Long-billed. Newport Beach, California, USA, Nov 1998 (Joe Fuhrman).

82.12 Juvenile Short-billed Dowitcher (*griseus*). Juveniles much brighter than nonbreeding adults, with heavily patterned upperparts, most feathers with internal as well as fringing buff to rufous. Breast bright buff at first, fading over time. Probably but not definitely this subspecies. Florida, USA, Sep 2002 (Richard Chandler).

82.13 Juvenile Short-billed Dowitcher (*hendersoni*). Like previous individual but very likely midcontinent subspecies, which is especially richly colored. Conneaut, Ohio, USA, Aug 2002 (Robert Royse).

82.14 Juvenile Short-billed Dowitcher (*caurinus*). Like the two more easterly subspecies but markings of upperparts tend to be darker and narrower. Profuse internal markings on tertials and generally brighter coloration distinguish from Long-billed. Ventura, California, USA, Sep 1999 (Brian Small).

82.15 Juvenile Short-billed Dowitcher (*caurinus*). At dull end of variation in Short-billed juveniles, could even be brightly marked Long-billed, some of which have internal tertial and scapular markings. Identified as Short-billed by relatively long primary projection, more characteristic of that species. Foraging with adult Long-billed. Bright green legs from adhering algae. Henderson, Nevada, USA, Aug 2003 (David Roemer).

82.16 Nonbreeding Short-billed Dowitcher (*griseus*). Long-billed sandpiper with tail looking paler than wings, white stripe up back; narrow white rear edge of inner wing also characteristic. Uniform coloration of mantle and not especially wide black tail bars both indicate this species. Naples, Florida, USA, Feb 1999 (Wayne Richardson).

83 Long-billed Dowitcher

(Limnodromus scolopaceus)

Still another bird of Beringia, this species is similar in every way to the Short-billed but focuses more on freshwater, where its longer bill may suit its tendency to wade belly-deep. Field identification is facilitated by the very different calls of the two species.

Size Weight 105 g, length 28 cm (11 in), male bill 6.2 cm, female bill 7.2 cm, tarsus 3.8 cm.

Plumages Slight size dimorphism, only evident in bill length (female's longer). Bill black, may be tinged gray or greenish at base; legs light green. *Breeding adult* with upperparts mostly black, feathers with fine jagged rufous crossbars and buffy to whitish fringes. Loral and short postocular stripe dark, fairly prominent supercilium rufous or partially white. Underparts entirely rufous (frosted with white in fresh plumage), sides extensively barred. *Nonbreeding adult* with bill greenish gray at base. Upperparts and breast brownish gray, sides with coarse gray bars. Distinct white supercilium, dark loral stripe, and short, dark postocular stripe. Posterior sides and undertail coverts spotted and barred with black. *Juvenile* with bill more extensively pale greenish gray at base than in adults. Distinguished from nonbreeding adult by pale rufous fringes on mantle, scapulars, coverts, and tertials and buff wash on underparts; otherwise very similar.

Identification Very similar to Short-billed Dowitcher in all plumages. Bill averages longer, and female bill distinctly longer than in Short-billed, but male Long-billed indistinguishable from either sex of Short-billed. Tail pattern different, Long-billed with dark bars always broader than pale ones, Short-billed with pale bars broader in interior populations and sometimes mottled rather than barred, but species overlap and often difficult to see tail bars clearly. Atlantic coastal populations tend to have wider dark bars, much like Long-billed. Primary projection slightly greater in Short-billed (typically two primary tips exposed in Short-billed, none in Long-billed) but may vary with subspecies. In *breeding* plumage, Long-billed typically with dark rufous bars and white tips on scapulars, forming more contrast than typical in Short-billed, with lighter rufous bars and buff tips. Some Short-billeds have white tips, however. Also, Long-billed central tail feathers newly molted with reddish ground color (same color as underparts when fresh, somewhat faded in fall), All three subspecies of Short-billed usually with white ground color whether newly molted or not (rarely with reddish tinge, uppertail coverts often reddish). Best distant field mark shown by black markings on upper sides below bend of wing, typically spots in Short-billed and bars in Long-billed, but occasional Short-billed looks quite barred. Farther back on sides, both species barred. Long-billed also averages slightly darker below and less likely to have white on belly except for narrow white tips to many feathers when in fresh plumage; when Short-billed shows white, it is more extensive (individuals in molt confuse this). Some of these characters of less value on worn individuals in fall. Some

Short-billeds distinctive in having spots scattered over entire underparts, other Short-billeds and all Long-billeds less spotted (but difficult to see in field). Both spring and fall molts in Long-billed lag slightly behind that of Short-billed, providing some help in field identification. Also, many if not all adult Long-billeds on their way south pause in migration to molt both body and wing feathers before continuing southward, and they remain at high latitudes much later than Short-billed. *Nonbreeding* plumage most similar, with minor but fairly consistent differences. Long-billed typically with plain dark gray breast, Short-billed with gray somewhat paler and less extensive and with scattered dark speckles; easier to see in museum than in field! Long-billed mantle feathers with distinctly darker centers, producing somewhat blotched effect; Short-billed feathers more uniformly gray, producing smooth gray-brown effect. Finally, barring on sides of Long-billed rather broad and even, same color as breast and often on pale gray background, barring on Short-billed somewhat narrower and more irregular, more contrasty on white background and usually darker than breast color. *Juveniles* also very similar but this plumage may provide most positive distinction between species. Long-billed a bit darker above and grayer on neck and breast, overall less brightly patterned. Short-billed more colorful, with conspicuous markings on scapulars and usually buffy neck and breast, Long-billed duller in these areas, with pale markings dull reddish rather than bright buff. Best distinction provided by tertials, dark and fairly heavily marked with pale lines in Short-billed, both fringing and internal. Same feathers slightly paler in Long-billed, with only fine pale fringe and typically no internal markings; should be no surprise to learn that occasional Long-billeds have internal markings. Juvenile Long-billeds begin body molt in southward migration, not usually so in Short-billed. See Short-billed Dowitcher for distinctions from other shorebirds.

In Flight Exactly like Short-billed, no obvious distinction. From below differs from snipes in having pale underwings and less contrast between breast and belly.

Voice Flight call a single or repeated *peep* or *keek*, when flushed often a rapid, twittered series. Much more likely to call than Short-billed, both in flight and while feeding. Breeding-ground song a series of buzzy notes ending in *ti-ti—weeburr*, ending sounding something like *dowitcher*.

Behavior Exactly like that of Short-billed except for greater propensity to use freshwater habitats. Probably never attains flock size of more common relative. Two dowitcher species flock together regularly but not universally, perhaps just because of some habitat and seasonal separation.

Habitat Breeders on wet tundra. Migrants and winterers in fresh and salt marshes and mudflats.

Range Breeds on arctic coasts of western and northern Alaska, Yukon, and Nunavut; also quite far west along north coast of Siberia. Winters on coasts from British Columbia and North Carolina south and across far southern USA south to Isthmus of Tehuantepec, less commonly farther south and through West Indies and rarely to northern South America. Migrants mostly through West in spring, both west and east of Rocky Mountains; more widespread in fall, especially juveniles, which reach New England coast in their rather late migration.

83.1 Breeding Short-billed and Long-billed dowitchers. Postbreeding individuals, Short-billed (in front) more worn than Long-billed. Note spots on sides of breast in former, bars in latter. Difference in posture when feeding sometimes evident, Short-billed with flatter back, Long-billed with rounded back; significance and cause unknown. Seattle, Washington, USA, Jul 2003 (Stuart MacKay).

83.2 Breeding Long-billed Dowitcher. Typical individual in fresh plumage, distinguished from Short-billed by bars on side of breast, entirely rufous underparts with no central spots, and whitish fringes on some belly feathers. Contrast between rufous and white edges on scapulars and coverts typical of this species; Short-billed usually more uniformly edged with buff or light rufous. Also note reddish central tail feathers. Upper Coast, Texas, USA, May 1998 (Brian Small).

83.3 Breeding Long-billed Dowitcher. Long bars on sides of breast and cinnamon central tail feathers both indicate species. Dowitcher underwings are more heavily marked than in most small sandpipers but less heavily than in snipes. Groups of wing coverts and axillars readily distinguished in this bird. Ventura, California, USA, May 1992 (Larry Sansone).

83.4 Breeding Long-billed Dowitcher. Worn post-breeding individual, usually not as worn looking as Short-billed when arriving in southbound migration. Long bill indicates female, longer than in any Short-billed. Seattle, Washington, USA, Jul 2003 (Stuart MacKay).

83.5 Breeding Long-billed Dowitcher. Later individual than previous one, more worn and, oddly, showing no bars on sides of breast. Lack of primary projection indicative of species, definitive call notes heard when flushed. Note how feeding area (algal soup, in this case) can affect leg color. Henderson, Nevada, USA, Aug 2003 (David Roemer).

83.6 Long-billed Dowitcher. Early stage of spring molt; note new scapulars typical of Long-billed, few bars on sides of breast, and tail feathers with much wider black than white bars, all indicative of this species. Newport Beach, California, USA, Apr 2001 (Joe Fuhrman).

83.7 Nonbreeding Long-billed Dowitcher. Plain in winter, very similar to Short-billed but mantle feathers with darker centers, unspeckled breast, virtually no primary projection, and black tail bars quite wide. Shortish bill indicates male. Ventura, California, USA, Oct 1986 (Brian Small).

83.8 Juvenile Long-billed Dowitcher. Much duller than typical juvenile Short-billed, closer to nonbreeding adult plumage but scapulars darker with buff fringes, often narrow fringes on tertials as well and buff wash on breast. Ventura, California, USA, Sep 1999 (Brian Small).

83.9 Long-billed Dowitcher. Just like Short-billed Dowitcher; calls should be heard for identification. Tail averages darker in this species but probably not usable as field mark. Salton Sea, California, USA, Dec 1989 (Mike Danzenbaker).

Snipes (Gallinagini)

The 19 species of snipes are denizens of freshwater marshes and bogs all over the world. Some of them are rather rare and poorly known residents with restricted ranges. Relatively solitary by nature and well camouflaged among marsh vegetation by their vivid stripes, they feed by probing the soil with their long bills. Most species have spectacular aerial displays during which they make haunting whistles or weird mechanical sounds. Four species occur in North America, but two of them are accidentals.

84 Jack Snipe

(Lymnocryptes minimus)

This little sandpiper could have been called "rail-snipe" because of its relatively short bill and skulking habits. It is also unusual in the vividness and iridescence of its back pattern, like camouflage taken too far. Accidental in North America and secretive, it is an unlikely find.

Size Weight 50 g, length 20 cm (8 in), bill 4.0 cm, tarsus 2.4 cm.

Plumages No obvious plumage variation. Bill dull pinkish with black tip; legs dull green.

Identification Half the bulk of other snipes, with considerably shorter bill, evident at rest and in flight. Head strongly patterned, with split supercilium but without pale crown stripe of other snipes. Back with broad yellowish stripes even more striking than in other snipes. Mantle feathers and scapulars with unique iridescent greenish glossed fringes. Breast and sides streaked, no indication of barring below as in other snipes.

In Flight Wings look shorter, more rounded than in other snipes, tail all dark and wedge shaped, bill shorter. Difficult to flush, and then soon drops back into marsh, more like rail than snipe. Inner wing with fine white trailing edge.

Voice Flushes silently. Rarely heard flight call is low pitched and quiet, somewhat harsh but very different from loud calls of Wilson's and Common snipes.

Behavior Feeds like other snipes, moving slowly and probing vigorously in mud. Often bounces up and down while foraging, combining vertical movement with forward progress (occasional bobbing also in Wilson's and presumably other snipes).

Habitat Breeders in forest bogs. Nonbreeders in wet meadows and muddy pond and stream edges. Flies to wet fields to feed at night.

Range Breeds across northern Eurasia. Winters south to central Africa and Southeast Asia. Accidental in North America; records from Alaska, Washington, California, Labrador, and Barbados, primarily in winter.

84.1 Jack Snipe. Very small snipe, dark with paired vivid buff stripes down head and back, also narrower but still conspicuous pair of stripes on upper scapulars. Short bill compared with other snipes, pointed dark tail. Kaarina, Finland, Feb 1997 (Henry Lehto).

85 Wilson's Snipe

(Gallinago delicata)

"Snipes are striped" is one of the best mnemonic devices for shorebird identification. Snipes are common in a habitat not much used by other shorebirds but shared by rails and bitterns. Ubiquitous all across North America in breeding or nonbreeding season or both, this species is common but not always as evident as other shorebirds. Overtly displaying breeding birds convey a very different impression than the skulkers of winter.

Size Weight 97 g, length 27 cm (10.5 in), bill 6.4 cm, tarsus 3.1 cm.

Plumages Bill black at tip, grading into grayish base; legs gray to greenish. Dark brown above, white below, with vividly striped head and back and barred sides. *Juvenile* with coverts averaging more richly colored than those of adults, bases darker brown and tips brighter buff to rufous. Adults usually with dark brown shaft streak extending to tip of each median and lesser secondary covert, juveniles without this streak but instead with faint black margin to same feathers. Probably not distinguishable in field.

Identification No other shorebirds as vividly striped above as snipes. Long bill on shorebird of this size only duplicated by two species of dowitchers, which lack conspicuous stripes on head and back and have gray, buff, or reddish breast depending on plumage, whereas snipe has striped and barred brown breast. See Common and Pin-tailed snipes, of peripheral occurrence in North America (Common may breed in western Aleutians). Vagrancy to North America by Great Snipe (*Gallinago media*) of Europe and western Siberia seems not impossible. That species would be recognized by its slightly larger size, more extensive pale coloration on bill base, prominently white-tipped wing coverts, extensively chevroned underparts, and conspicuously white outer tail feathers.

In Flight Medium-small sandpiper with long bill and quite different appearance from most sandpipers. Dark brown above and on both surfaces of wings, white belly. Dark underwings distinguish from dowitchers in flight. At close range, striped head and back and orange tail markings may be visible. Solitary; often flushes from dense marsh vegetation. Typically flies straight up into air ("towering") and rapidly departs area or circles around before landing nearby.

Voice Flight call a loud *scaip* (could be construed as *snipe*), typically given only once or twice as bird flies away. Less often a bird flying over will call. Breeding-ground "song" a loud, continued series of notes (*chip-er-chip-er-chip*, etc.). Aerial display on breeding grounds involves "winnowing," rising and falling musical sound (*whi whi whi whi WHI WHI WHI WHI whi whi whi whi*, etc.) given by male. Produced by wing beats pushing air through widespread outermost tail feathers as bird ascends high in air and then drops, gaining speed on each dive.

Behavior Snipes stitch wet mud and shallow water as enthusiastically as dowitchers but move even more slowly. Perch conspicuously in open during breeding season.

Habitat Freshwater marshes throughout year, moving to salt marshes and flooded fields during nonbreeding. Feeds in all kinds of shallow marshy environments, some with dense vegetation and some quite open.

Range Breeds widely across North America from Alaska to Newfoundland and south to California, Colorado, Great Lakes, and New England. Winters mostly south of breeding range, from British Columbia, Nebraska, and New York south to West Indies and northern South America. Migrants across continent.

85.1 Wilson's Snipe. Unmistakable heavily striped, long-billed sandpiper of marshy areas. Dark and light stripes on sides of head, heavy barring on sides of body. Much longer billed than any striped calidridine and more heavily striped than dowitchers and woodcocks with similarly long bills. Any other snipe only a casual visitor to North America. Conneaut, Ohio, USA, Aug 2003 (Robert Royse).

85.2 Wilson's Snipe. Brightly colored tail well shown, with outermost feather black and white barred; showing no hint of orange is good distinction from Common Snipe. Note mantle and scapular lines broken when feathers fluffed. Salmon Arm, British Columbia, Canada, Aug 2003 (Syd Roberts).

85.3 Wilson's Snipe. Long-billed sandpiper with brown breast, white belly, dark underwings (distinction from Common Snipe). Note tail extends beyond toes (distinction from Pin-tailed Snipe). Weld County, Colorado, Jan 2004 (Bill Schmoker).

86 Common Snipe

(Gallinago gallinago)

Common Snipe is not likely to be identified in North America except in the hand or on the western Bering Sea islands, where it replaces Wilson's Snipe. It may be more regular and widespread on this continent than we will ever know; the checking of hunters' bags would surely be a good way to find out.

Size Weight 110 g, length 27 cm (10.5 in), bill 6.7 cm, tarsus 3.2 cm.

Plumages No obvious plumage variation.

Identification Looks exactly like Wilson's Snipe in field. Has been called "warmer," with back slightly paler and stripes with buffier cast than Wilson's, but the two overlap in this as in just about all characters. Claimed in literature that brown barring on sides on light brown ground color less contrasting than darker brown barring on white in Wilson's, but specimens of Common from Russia—source of North American records—identical to Wilson's. Tertials in Common usually strongly barred where visible; in Wilson's, bars usually become more obscure toward base, but again overlap in this character. Wings and tail furnish best characters. Common extensively white on underwing coverts, Wilson's mostly dark with only narrow white barring; however, some juvenile Commons almost as dark as Wilson's. Axillars barred dark and light, dark bars same width as or narrower than light in Common, wider than light in Wilson's. Secondaries tipped with white in both species, but tips average considerably wider in Common, forming square-cut patch, whereas white narrower in Wilson's and extending around as fringe at tip; some juvenile Commons almost as narrow as Wilson's. Fourteen tail feathers in Common to Wilson's 16, but this of little help in field. Tails also differ in color of outermost feather—barred black and white in Wilson's, with orange on inner web in Common—but this might be as difficult to see clearly as number of feathers! Wait for a wing and tail stretch for a documenting photo.

In Flight Perhaps best identified in flight but still a difficult call. Could only be distinguished by more white in underwing coverts and broader white tips to secondaries (very close in the two species and might just about overlap), but snipes in flight are elusive targets for binoculars. Good photo with stroke of luck might confirm fleeting impression. Male snipe in flight display with tail widespread should be readily identifiable, however. Common has one rather broad feather on each side separated from other feathers, Wilson's usually has two distinctly narrower feathers isolated on either side.

Voice Flight call and song much like that of Wilson's, but winnowing sound very different, a deeper, hollower, even more electronic sound.

Behavior No behavioral difference documented between Wilson's and Common snipes.

Habitat As far as known, identical to that of Wilson's.

Range Breeds across Eurasia. Winters south to Africa and Southeast Asia. May breed in western Aleutians, numerous specimens in migration from Bering Sea islands; also records from Labrador.

86.1 Common Snipe. Basically identical to Wilson's Snipe when perched; best distinguishing characters in wings and tail, not usually visible. Salo, Finland, Aug 1999 (Henry Lehto).

86.2 Common Snipe. No apparent geographic variation from Europe to Siberia. Numerous records in Alaska, could be more regular but very difficult to detect. Kyoto Prefecture, Japan, Aug 2000 (Tomokazu Tanigawa).

86.3 Common Snipe. Underwing view allows identification, showing typical pale underwing coverts, mostly white axillars, and broad white tips to secondaries. Some juvenile Commons darker, but Wilson's never shows so much white. Kyoto Prefecture, Japan, Apr 2000 (Tomokazu Tanigawa).

86.4 Common Snipe. Male during flight display with tail spread with one broad feather isolated on either side of tail. Wilson's has two narrower feathers showing on either side, Pin-tailed a series of very narrow feathers. Underwing barred but not as dark as in Wilson's or Pin-tailed. Myvatn, Iceland, Apr 2003 (Göran Ekström).

86.5 Adult snipe wing specimens. Common (below) with more white in underwing coverts and axillars and broad, square tips on secondaries. Wilson's Snipe, Bellingham, Washington, USA, Oct 1995; Common Snipe, Kamchatka, Russia, Jul 1992.

87 Pin-tailed Snipe

(Gallinago stenura)

This common Asian species complicates snipe identification in North America's farthest West, where it is accidental, but it is a bit more different from Common and Wilson's snipes than they are from one another. On its breeding grounds, its aerial displays include memorable, almost electronic, sounds made by the highly modified tail.

Size Weight 115 g, length 27 cm (10.5 in), bill 6.2 cm, tarsus 3.3 cm.

Plumages No obvious plumage variation.

Identification Much like Wilson's and Common snipes but slightly shorter- and thicker-billed and with more rounded wings. Tail does not project beyond wings, shorter wings and shorter tail produce blunter-looking rear end. Tail with 28 feathers, far above mode of 12 in most birds; outer ones white and very narrow, hence name. This diagnostic mark could be seen in bird stretching one wing and tail, but it would have to be relaxed, not the object of attention of excited birders (with digiscoping techniques of today, photography of relaxed shorebirds more frequent). Overall looks browner and slightly less contrastingly marked than Wilson's and Common. Face looks paler than Wilson's or Common because pale supercilium distinctly wider than dark loral stripe and loral stripe slightly narrower and paler; in Common and Wilson's, stripes closer to equal (pale slightly wider than dark) and loral stripe more distinct. White or pale buff on outer edge of lower scapulars extends onto inner edge in Pintail, whereas in Wilson's and Common pale edging interrupted short of tip, buff to reddish when renewed on inner edge; thus Pin-tailed looks scalloped as much as striped. Pale fringes on median and lesser wing coverts typically broader than in Common and Wilson's, so coverts overall paler but also more patterned. Swinhoe's Snipe (*Gallinago megala*) of eastern Asia another possible vagrant to North America; so similar to Pin-tailed that field identification problematic. Averages larger, longer-winged, and longer-tailed than Pin-tailed, with thicker legs and somewhat darker in general, but definitive field marks not discovered. Swinhoe's usually has 20 tail feathers, outermost ones wider than in Pin-tailed; probably only distinguishable in hand.

In Flight Paler upperwing coverts contrast with rest of wing; good mark if it can be seen. Juvenile with slightly narrower buff tips to coverts, so a bit more similar to other species. Underwing dark like Wilson's rather than white-marked like Common, also trailing edge of wing without obvious white secondary tips as in Common. Less likely than Wilson's and Common to zigzag and fly high and away.

Voice Flight call sounds slurred; lower pitched and harsher, more "throaty" than Wilson's and Common snipes. Winnowing of aerial display very different from that of other snipes, a series of cricketlike single notes that accelerate into a rattle and then a sizzling sound as bird pulls out of dive.

Behavior Much like that of other snipes. Sometimes flies in mixed groups with Common.

Habitat Breeds in open and shrub tundra and wet meadows, somewhat drier areas than Common Snipe. Winters on freshwater shorelines and in marshes, wet meadows, and flooded fields.

Range Breeds across Siberia and central Asia. Winters from India through Southeast Asia. Accidental in North America, two spring records from western Aleutians.

87.1 Pin-tailed Snipe. Distinguishable from Common and Wilson's snipes by shape (shorter wings and tail) and subtle features of coloration. Pale supercilium broadens even more in front of eyes than in other species, loral stripe quite narrow in comparison. White fringes on scapulars extend around inner side prominently; in other species usually replaced by dull buff. Coverts prominently white tipped, giving more scalloped appearance to closed wing. Kyoto Prefecture, Japan, Sep 1998 (Tomokazu Tanigawa).

87.2 Pin-tailed Snipe. In flight, lesser and median coverts distinctly paler than rest of wing, contrast greater than in other two species. Underwing and secondaries as in Wilson's, not Common. Beidahe, China, Aug 2002 (Brent Stephenson).

87.3 Pin-tailed Snipe. Species is easily distinguished in hand by numerous narrow outer tail feathers. This could be seen in a bird stretching wing and tail. Black, orange, and white color of tail typical of all three *Gallinago* snipes, but outer feathers distinctive in each species. Anne Plains, Australia, Apr 1994 (Clive Minton).

Woodcocks (Scolopacini)

The six species of woodcocks are among the most aberrant shorebirds, never going near the shore but remaining in forested and shrubby habitats of the Northern Hemisphere. Even more solitary than snipes, they feed like them, probing the soil with super-long bills. Their dead-leaf pattern adapts them perfectly to the scattered light on the forest floor. One species is common and widespread in North America, and another is very rare.

88 Eurasian Woodcock

(Scolopax rusticola)

This woodcock is very similar to its American relative in habits, and the most likely birders to detect it if it does visit this continent again are those not on the prowl for shorebirds. Its breeding range is almost entirely north of the American species, so it should be a more likely candidate for vagrancy.

Size Weight 310 g, length 33 cm (13 in), bill 7.2 cm, tarsus 3.6 cm.

Plumages All plumages much alike, slightly different pattern on primary coverts of juvenile impossible to see in field.

Identification Typical woodcock, with earth brown plumage, big head and eyes, very long bill, short tail, rounded wings, and woodland/edge habitat. Hindcrown conspicuously barred and back conspicuously striped as in American Woodcock. Differs from that species in considerably larger size, heavily barred brown and white underparts, barred instead of gray neck and lesser wing coverts, and less contrasty mantle and scapular lines.

In Flight Distinctly more pointed wings than American Woodcock, larger and more round-winged than snipes. Outer web of all primaries barred, unbarred in American Woodcock.

Voice Silent when flushed. Rarely gives snipelike call.

Behavior Much like American Woodcock in all habits. Feeds at night, also during day, by both picking and probing, turning over leaves to expose prey.

Habitat Woodland with dense undergrowth for breeding. Similar in winter and migration but ranges more into shrubby habitats.

Range Breeds across Eurasia. Winters south to North Africa and Southeast Asia. Vagrant to northeastern North America; more than five winter records, but almost all very old, as far west as Quebec, Pennsylvania, and Alabama.

88.1 Eurasian Woodcock. Heavily barred underparts, lack of gray stripes above, and larger size distinguish this very rare visitor from American Woodcock. Chita, Aichi Prefecture, Japan, Jan 1987 (Tadao Shimba).

89 American Woodcock

(Scolopax minor)

The only forest-based shorebirds, woodcocks are grouselike in their plump shape and woodland colors, but their long bills offer easy distinction. They are seldom seen, yet this species has been considered the most abundant shorebird of North America because of the two million taken annually by hunters. Recent population declines may be caused by diminishing second-growth forests.

Size Male weight 149 g, female weight 186 g; length 30.5 cm (12 in), bill 6.8 cm, tarsus 3.5 cm.

Plumages Moderate size dimorphism (female larger) but unlikely to be opportunity for comparison. Bill dull pinkish, extreme tip black; legs dull pinkish. *Adult* mottled black, pale brown, and cinnamon above with conspicuous pale mantle stripes, mostly cinnamon below. Crown crossed by conspicuous wide black bars. *Juvenile* when just fledged with throat and foreneck same gray-brown as hindneck, contrasting more with white chin than in adult, but later in season looks exactly like adult in field and might be distinguished only by tertials, less worn in early fall and more worn in winter.

Identification No other North American shorebird anything like this species; easily distinguishable from snipes by pattern and habitat. See also very rare Eurasian Woodcock.

In Flight Earth brown above and rusty below, with rounded wings, short tail, and very long bill held pointing down. Flushes explosively, and in flight looks more like very small grouse than medium-small sandpiper; fortunately, there are no long-billed grouse. Rapid and twisting flight and dim lighting in its habitat make complex pattern of upperparts unlikely to be seen.

Voice No vocalization when flushed, but male has somewhat nighthawk-like *peent* call from ground during breeding season. Narrowed and shortened outer primaries presumably responsible for whistling or twittering sound made when flushed and during courtship flight (male's outer primaries slightly narrower than female's).

Behavior Solitary. Foraging consists of probing deeply in soil for earthworms and other invertebrates, both day and night (latter more on wintering grounds). Usually seen in daytime when flushed from ground, quickly disappearing into woodland. Moves into open areas at night for feeding. Breeding male has spectacular display on ground and impressive display flight at dusk and dawn. Displaying males may be seen in steady flight over open areas and forest canopy.

Habitat Breeders in forest, especially earlier successional stages. Nonbreeders more commonly in later forest stages, commonly bottomland hardwood forests; dense understory preferred.

Range Breeds across deciduous forest zone of eastern North America, from Manitoba and Newfoundland south almost to Gulf coast and into northern Florida and west to eastern edge of Great Plains. Winters across Southeast from Missouri and New Jersey south, except southern Florida. Migrants throughout East but not obvious like other shorebirds; early in spring and late in fall. Vagrant west to Continental Divide, rarely farther west, all the way to California, and south to Tamaulipas and Quintana Roo.

89.1 American Woodcock. Long-billed and short-legged woodland sandpiper, grouse colored to match usual habitat. Black crossbars on head and gray mantle and scapular stripes offer further field marks, but nothing else like it except very rare Eurasian Woodcock. Round winged and short tailed in flight, seen usually when flushed or displaying. Chicago, Illinois, USA, Mar 2003 (Kanae Hirabayashi).

Phalaropes (Phalaropodini)

Phalaropes, at least two of them, are as rightly called seabirds as shorebirds, as they spend over half their life at sea, often well out of sight of land. They are modified for swimming, with short legs and fleshy lobes on their toes like little coots or grebes. They pick tiny invertebrates from the surface, and they often feed by spinning, creating a current that draws their prey toward them. Wilson's is a good swimmer but winters on alkaline lakes instead of oceans and forages on shore much more often than the other two. All, however, exhibit the unusual phenomenon of reversed sexual dimorphism, females larger and more colorful than males and taking the lead in courtship, while males build nests, incubate eggs, and care for young. All three species occur in North America.

90 Wilson's Phalarope

(Phalaropus tricolor)

This species is characteristic of marshy puddles over a wide part of the western interior, where it associates with avocets and stilts. It gathers in huge flocks to molt and fatten at certain saline lakes in the interior (Mono Lake the best known), then proceeds south to winter at similar spots in the Andes, where its companions are different kinds of avocets and stilts as well as flamingos. The chestnut, rufous, gray, and white females are models of neat beauty and elegance.

Size Male weight 51 g, female weight 68 g; length 22 cm (8.5 in), bill 3.3 cm, tarsus 3.3 cm.

Plumages Slight dimorphism in size and moderate dimorphism in plumage, females larger (not that obvious in field) and brighter. Bill black, legs brownish gray to black. *Breeding adult* in most cases brightly marked with black and rufous neck stripe and orange foreneck. Female plain gray (including crown) and rufous above, neck stripe vivid, extending as narrow chestnut stripe along lower scapulars. Male more variable, brightest ones superficially like females but differ by dark gray crown (best characteristic for sexual distinction), longer white supercilium (extending behind eye), and more mottled upperparts, feathers with darker centers than those of females. Many males pallid versions of this, grayish brown with little indication of reddish. Dullest males differ from nonbreeding plumage primarily by leg color and darker, mottled back. Body and partial flight-feather molt accomplished at staging areas in North America on southward flight. *Nonbreeding adult* with legs olive to yellow or orange-yellow, changing from breeding color in early fall and back into breeding color in late winter, before migration. Plain gray above and white below, with white supercilium and darker gray postocular stripe. Scapulars and tertials narrowly fringed with white, which wears off by late winter. *Juvenile* with legs as in nonbreeding adult. Hindneck gray, crown and rest of upperparts brown with all feathers buff fringed, creating strongly scalloped or, in some individuals, striped (especially on mantle) pattern. Fairly distinct dark eye stripe and white supercilium. Fresh-plumaged individuals with buff-washed breast. Body molt rapid while southbound through North America.

Identification Behavior distinguishes it as phalarope. In breeding plumage most similar to Red-necked Phalarope but larger and paler, with mostly light rather than dark head. Nonbreeding adult pale above and pure white below but lacks dark "phalarope mark" of other two species. Somewhat like Stilt Sandpiper and Lesser Yellowlegs, and at times feeding with them, but with whiter breast, more needlelike bill, and shorter legs. Other pale-breasted, long-billed sandpipers have differently shaped bills and often longer legs and forage differently. With long neck and long, slender bill, very different from similarly pale sandpipers such as Sanderling. Colored rather like nonbreeding Marsh Sandpiper but with shorter bill, much shorter legs, different behavior.

In Flight Medium-small sandpiper with plain wings and white rump and tail; gray-brown above, white below, with colorful head and neck pattern in breeding season. Grayer, less brown, than yellowlegs and Stilt Sandpiper with similar flight pattern; legs distinctly shorter, foot projection only part of toes. Underwings whiter than in similar species.

Voice Mostly silent during migration and winter. Both sexes give nasal grunting call (*ernt*) on breeding grounds which may be used as contact call, given especially frequently by courting females. Grunting call also heard in migration but does not carry very far.

Behavior Typical phalarope, spending much of time feeding on water, picking prey from surface and spinning like other two species, but quite different from other two in commonly wading in shallow water or even feeding along shore (toes not lobed as in other phalaropes). Often forages near larger species such as ducks and avocets which stir up water. Looks like phalarope even when foraging on land, with characteristic long neck and plump body, short legs, erratic gait, and different posture than most shorebirds (crouches with head forward). Only rarely goes to coasts but migrates and winters on inland water bodies, especially saline ones. When at coast, likely to be on shore of protected estuary.

Habitat Breeders in freshwater marshes and lake edges. Winterers on alkaline and saline lakes, often in mountains. Migrants mostly on similar lakes.

Range Breeds in western interior from British Columbia and Manitoba south to California and Colorado, sparingly east to southern Great Lakes and rarely to Atlantic coast. Winters in southern South America (small numbers north to southern Texas and West Indies). Migrants common throughout much of West and Great Plains, sparingly to Pacific coast from Canada to Oregon and throughout East, where more common southward. Locally common in Middle America, rare in West Indies.

90.1 Breeding female Wilson's Phalarope. Phalaropes exhibit reversed sexual dimorphism, with larger, more brightly colored females. This one slender billed, chestnut and gray above, with black neck stripe and orange breast. Legs dark in breeding season. Benton Lake, Montana, USA, Jun 2002 (Brian Small).

90.2 Breeding male Wilson's Phalarope. Duller, sometimes much duller, version of female, with crown dark, back mostly brown, enough of characteristic head pattern visible for identification. Benton Lake, Montana, USA, Jun 2002 (Brian Small).

90.3 Breeding male Wilson's Phalarope. Duller male; some show no bright coloration at all. Burns, Oregon, USA, Jun 2000 (Netta Smith).

90.4 Nonbreeding Wilson's Phalarope. Plain gray above, white below; this individual immature, indicated by retained white-fringed coverts and tertials; adult would have tertials gray like scapulars. Legs yellow in nonbreeding birds of all ages. Always associated with water but frequently feeds on land, where it forages with different posture than other sandpipers. Ventura, California, USA, Sep 1998 (Brian Small).

90.5 Juvenile Wilson's Phalarope. Above dark with buff fringes on all feathers, narrow line through eye, rich buff on breast that rapidly fades. Ventura, California, USA, Aug 1998 (Brian Small).

90.6 Juvenile Wilson's Phalarope. Typically faded and worn older juvenile, soon to molt. Toes not especially lobed in this species. Mono Lake, California, USA, Aug 1989 (Larry Sansone).

90.7 Breeding Wilson's Phalarope. Flight pattern plain wings and white tail, same appearance as Lesser Yellowlegs and Stilt Sandpiper but a bit smaller, with much shorter legs. Unmistakable in breeding plumage; three left birds females, others males. Arapaho National Wildlife Refuge, Colorado, USA, May 1992 (Erik Breden).

91 Red-necked Phalarope

(Phalaropus lobatus)

This, one of the tiniest seabirds (only a few storm-petrels are smaller), is at home on the broad ocean surface but may be encountered as well in large numbers in inland waters in migration. Although difficult to assess the population size of a bird that breeds all across the uninhabited tundra and winters at sea, it may be declining. Important staging areas in the North Atlantic have been abandoned in recent years, which could be from changing oceanographic conditions, as the species is still abundant in the North Pacific.

Size Weight 35 g, length 18 cm (7.25 in), bill 2.2 cm, tarsus 2.0 cm.

Plumages Slight plumage dimorphism, females brighter; females also average 6 g more than males in weight. Sexes more similar than in other two phalaropes in size and color. Bill black, legs dark gray to black. *Breeding adult* dark slaty gray above and on sides of breast, white below with gray stripes on sides. Broad rich buff mantle and scapular stripes. Much of head black, throat white and sides of neck rufous, either extending across breast or split by gray of breast. Male usually obviously duller and more heavily patterned than female. Rufous neck patch varies from about as large and bright as in typical female to virtually lacking, replaced by gray. Best sexual distinction better-developed white supercilium in male, generally white before and over eye and rufous behind it, joining rufous neck patch (white eye patch and rufous neck patch usually separated in female). Tertials and, in some individuals, scapulars more complexly patterned than in females. Pattern becomes simpler in males by midsummer, when many pale fringes worn off. *Nonbreeding adult* gray above, with fine white fringes to long, pointed scapulars and tertials giving striped effect, and white below. Hindcrown and "phalarope mark" (combination postocular stripe and auricular spot) black. White edges on scapulars and tertials wear off sufficiently so stripes less evident by midwinter. Tertials molt fairly late, dull brownish and contrasting with fresh gray scapulars until late in fall. *Juvenile* with head as in nonbreeding adult, but dark coloration more extensive on crown. Mantle feathers and scapulars fringed with bright buff or gold which fades to white; tertials white edged. Neck and breast with prominent buff or rusty wash on young juveniles which fades to white with age. Juvenile phalaropes change appearance rather quickly in fall through fading and rapid molt into first-winter plumage. Even when gray backed, immatures distinguished from nonbreeding adults through most of fall by dark crown and dark, white-edged tertials.

Identification Phalaropes unlikely to be mistaken for nonphalaropes because of behavior. In breeding plumage, somewhat like Wilson's but smaller, darker, and striped above, with white throat patch more strongly set off from neck color; very different from Red. In nonbreeding plumage much like Red because of head pattern and overall coloration but distinctly smaller, with striping on upperparts apparent at closer range; quite similar at distance. Finer bill on Red-necked good mark if it can be seen. Juveniles rather similarly colored and go through same plumage changes; thin bill of Red-necked still good distinction between them. Also, Red-necked has much more conspicuous golden mantle stripes.

In Flight Small sandpiper with conspicuous white wing stripes; rump and tail with indications of *Calidris* pattern, central parts of rump and tail somewhat darker than white sides, tail darker than rump. Very white below, head and neck dark in breeding plumage. Smaller than very similar Red Phalarope, not quite so pale in winter and very different in breeding plumage; check for thinner bill. Phalaropes in flight over ocean not in cohesive flocks like other shorebirds.

Voice Flight calls simple and hard (*tic, tic*), given intermittently but distinctive. Lower pitched than those of Red Phalarope. Moderately noisy in migration. Breeding-ground calls repetitious chatter much like flight calls.

Behavior Almost always on water, only rarely running along shore, perhaps only when blown out of normal habitat. Singles to small flocks can be common on ocean, especially at convergence lines ("tide rips"), where tiny prey congregated in large numbers by currents flowing together. Phalaropes twirl around and around, creating miniature upwelling that draws planktonic invertebrates to surface to be picked off by fine bill. Often seen in flight, presumably searching for richer food sources. Usually on larger bodies of water, often saline or alkaline, when inland, but may turn up in pools and ditches, especially when blown in off ocean.

Habitat Breeders on tundra and muskeg ponds. Nonbreeders on open ocean, fresh-water lakes, and coastal bays.

Range Breeds across arctic and subarctic latitudes from Alaska to Newfoundland, wider latitudinal spread than many arctic shorebirds. Winters on southern oceans but north to Colima off Mexican Pacific coast. Migrants largely offshore but large numbers on Great Basin lakes, sparingly across continent. Common off Pacific coast of Middle America, rare on Caribbean coast and in West Indies.

91.1 Breeding female Red-necked Phalarope. Small phalarope with needlelike bill, only rarely feeds on land. All-dark head with rufous neck stripe and white throat. Females typically with only white spot above eye. Dark back with light stripes the opposite of Wilson's Phalarope pattern. Nome, Alaska, USA, Jun 1998 (Brian Small).

91.2 Breeding male Red-necked Phalarope. Males typically a bit duller than females, always with more white over eye, often forming complete supercilium. Nome, Alaska, USA, Jun 1998 (Brian Small).

91.3 Adult Red-necked Phalarope. Adults molt quickly into nonbreeding plumage during southbound migration, become patchy looking. Ventura, California, USA, Aug 1997 (Brian Small).

91.4 Nonbreeding Red-necked Phalarope. Back gray, much of head and underparts white. Distinguished from Red Phalarope by feather edges producing somewhat striped look, also smaller size and much more slender bill. "Phalarope mark" around eye characteristic of both species. Newport Beach, California, USA, Nov 1996 (Joe Fuhrman).

91.5 Juvenile Red-necked Phalarope. Young juveniles with bright buff stripes above, brown breast, vaguely reminiscent of breeding adult. Bright colors quickly fade during southward migration. Turku, Finland, Aug 1992 (Henry Lehto).

91.6 Juvenile Red-necked Phalarope. Not yet in molt but changed dramatically from its appearance at a younger age just by fading of buff and brown colors. Note lobes on toes. Ventura, California, USA, Sep 1997 (Brian Small).

91.7 Juvenile Red-necked Phalarope. Pattern much as in small *Calidris* sandpipers, with white wing stripe and dark stripe down center of tail. Dark back with white stripes indicates juvenile. "Phalarope mark" usually visible in nonbreeding plumages, and landing on water is sure sign of phalarope. Santa Clara County, California, USA, Sep 1990 (Mike Danzenbaker).

92 Red Phalarope

(*Phalaropus fulicarius*)

Most of the year, this most oceanic of phalaropes is to be sought well offshore, where it indulges in at least a little plankton-straining like the whales it sometimes accompanies. These sea-going birds are typical seabirds, nothing but black, gray, and white, giving them the name Grey Phalarope outside the Americas. Breeding females, however, are as bright as any shorebird.

Size Male weight 50 g, female weight 57 g, length 22 cm (8.5 in), bill 2.2 cm, tarsus 2.2 cm.

Plumages Moderate plumage and slight size dimorphism, females brighter and slightly larger. Bill yellow with black tip in female, yellow duller and less extensive in male; legs grayish olive to yellowish. *Breeding adult* typically with striking white cheek patches, all or mostly chestnut below. Upperparts medium gray, fringed by pale buff and white, forming conspicuous white stripes. Female usually all chestnut below, head pattern vivid black and white. Males very variable, some femalelike but with duller underparts, finer black and cinnamon stripes above, streaked brown crown, and less extensive and conspicuous face patch. Even most femalelike of males distinguished by brown- and black-streaked crown. White cheek patch varies from as conspicuous as in female (although usually narrower) to virtually absent, in some birds represented only by white supercilium. Back looks more finely striped than that of female, as all mantle feathers dark centered and buff edged. Pale stripes reddish buff, in contrast with yellow-buff stripes of female, and as in females, stripe on either side of mantle may be more conspicuous than others. Males typically more extensively white below than females, darkest individuals about like lightest-bellied females and lightest individuals largely white below. Many look largely chestnut because white belly not visible in water. Body and partial flight-feather molt accomplished at sea at high latitudes before continuing southward. *Nonbreeding adult* with bill black, extreme base usually pale. Upperparts entirely pale gull-gray, tertials darker gray narrowly fringed with white in fresh plumage, underparts white. Black patch on rear of head and black "phalarope mark," a broad line from eye over ear coverts. At least some individuals retain scattered dark red feathers on center of underparts very late in fall, not usually evident in field. *Juvenile* with bill entirely black. Much brighter than nonbreeding but very different from breeding. Black "phalarope mark" joined to black hindneck and much of crown; supercilium and throat white. Back finely streaked cinnamon and black like breeding male, tertials black with buff edges. Neck and breast buffy to reddish brown, rest of underparts white. Rusty fades quickly, whereas crown becomes mostly white by molt. Striped back feathers gradually replaced by gray in fall, except dark tertials which persist well into winter.

Identification Nothing much like it in breeding plumage, although full-plumaged Curlew Sandpiper somewhat similarly colored. Nonbreeding adults and juveniles easily mistaken for Red-necked Phalarope, but look for thicker bill in Red, distinctly larger size if comparison possible. Winter adult Red with entirely uniform gray back, Red-necked with distinct (although not conspicuous) stripes. Superficially like Sanderling and rarely feeds on beaches with that species but easily distinguished by "phalarope mark" and erratic gait with very short legs.

In Flight Brown above and chestnut below in breeding plumage, with dark cap and white cheeks; white underwings contrast with dark underparts. Combination of conspicuous wing stripe and dark upper- and underparts, and little or no white in rump or tail, distinctive. Pale gray above and white below in nonbreeding plumage; somewhat like *Calidris* sandpipers, with central parts of rump and tail darker than white sides, tail darker than rump. Nonbreeders and juveniles much like Red-necked Phalarope; distinguish by larger size (comparison often possible during migration) and thicker bill of Red. Nonbreeders also much like Sanderling but with paler greater primary coverts, so leading edge of wing not dark as in that species; also, white stripe distinctly narrower on outer wing (looks broader in Sanderling). Landing on water is almost sure sign of phalarope, but offshore occurrence is not sufficient clue, as Sanderlings and other shorebirds may be in migration far out over ocean.

Voice Rarely heard, a *pit* call similar to that of Red-necked Phalarope but slightly higher pitched and more musical. No flight display as such on breeding grounds but very vocal, including twittering calls and raspy *weep*, oft repeated.

Behavior Typically feeds in small flocks, even during breeding season. Most ocean-going of phalaropes and most likely to be seen well offshore, where singles and small flocks drift and spin for planktonic prey. Relatively broad bill has fine structures on edges like lamellae of ducks, providing straining apparatus not found in other two phalaropes, but most feeding by visual detection and picking from water. Usually on large bodies of water when seen inland. Either very late in migration or frequently wintering, sometimes comes ashore in large "wrecks" on Pacific coast in October–December, feeding in any sort of water body or even on beach, where birds run erratically in very different manner from other sandpipers. Apparently ocean surface *can* get too rough for them to forage successfully, and many birds perish. Beachcombers should be alert for chance to examine these pelagic shorebirds in hand.

Habitat Breeders on tundra ponds, nonbreeders mostly on open ocean, vagrants usually on large lakes.

Range Breeds across Arctic from northern Alaska to northern Quebec in much narrower latitudinal belt than Red-necked; also across northern Eurasia. Winters on tropical and southern oceans but mostly eastern Pacific and eastern Atlantic, locally off coasts from California south and North Carolina to Florida; present at least into midwinter regularly north to Washington. Migrants almost entirely offshore, rarely to anywhere on continent but recorded widely. Common off Pacific coast of Middle America, very rare on Caribbean coast and in West Indies.

92.1 Breeding female Red Phalarope. Most seagoing of phalaropes, with thick bill. A real knockout in this plumage, with yellow bill, black and white head, vividly striped back, and all-rufous underparts. Barrow, Alaska, USA, Jun 1989 (Erik Breden).

92.2 Breeding male Red Phalarope. Males always duller, crown usually streaked, face patch smaller and less distinct, and back striping finer, less dramatic. Some males much duller, not much brighter than nonbreeding plumage. Note exposed wing coverts, usually hidden by other feather tracts. Iceland, Jun 2000 (Thomas Roger).

92.3 Adult Red Phalarope. Rapid molt of mantle and scapulars to become gray seabird, tertials later. Reddish on neck indicates adult. Ventura County, California, USA, Sep 1971 (Brian Small).

92.4 Nonbreeding Red Phalarope. Gray above and white below, much like Red-necked Phalarope but no trace of striping on upperparts, bill obviously thicker (but mostly black in nonbreeding season). Usually at sea but driven inshore by storms. Monterey, California, USA, Nov 1996 (Joe Fuhrman).

92.5 Juvenile Red Phalarope. Colored much like juvenile Red-necked, but thicker bill distinctive. Brown on breast fades quickly to white during southbound migration. Barrow, Alaska, USA, Aug 1976 (Peter Connors/VIREO).

92.6 Immature Red Phalarope. Mantle and scapulars quickly molt into gray much like adult, but dark crown and tertials (latter retained through winter) indicate first-year bird. Note short legs and their placement for effective swimming. Lobed toes characteristic of this and Red-necked. Conneaut, Ohio, USA, Oct 2003 (Robert Royse).

92.7 Breeding female Red Phalarope. Very dark with vivid white wing stripe in this plumage; white face patch and white underwings contrasting with dark underparts diagnostic. Offshore Sonoma County, California, USA, May 1991 (Mike Danzenbaker).

92.8 Nonbreeding Red Phalarope. Gray back and broad white wing stripe; rump and tail pattern somewhat like small *Calidris* sandpipers. Distinguished from Sanderling by evenly colored wing; inner part paler than outer part in Sanderling (but dark lesser coverts contrast with paler median coverts), same as outer part in phalarope. Off Carmel, California, USA, Aug 2003 (Glen Tepke).

Pratincoles and Coursers (Glareolidae)

The 17 species of this Old World family include the odd Egyptian Plover, 8 coursers, and 8 pratincoles. Coursers are long-legged, short-winged, big-headed birds of deserts and bushlands that forage rather like plovers or dig for insects with curved bills. Pratincoles are long-winged, short-legged, fork-tailed birds of lake and river margins that forage for flying insects much like swallows do and are similarly gregarious. Their long wings have brought a single species to North America from each direction, both as accidentals.

93 Collared Pratincole

(Glareola pratincola)

Diverging from the usual picture of a shorebird, pratincoles justify the alternate name "swallow-plover." This European species is included here only because one set out one autumn day across the Atlantic Ocean and made landfall at the farthest southeast point of North America.

Size Weight 80 g, length 25 cm (10 in), bill 1.4 cm, tarsus 3.2 cm.

Plumages Bill black, red at extreme base, legs dark olive. *Breeding adult* plain medium brown above, with brown breast, buffy sides, and white belly, with conspicuous buff throat patch bordered by narrow black line. *Nonbreeding adult* with black throat border broken into fine streaks, bill base duller. *Juvenile* with bill base pale pink, pale cream throat, mottled brown breast, and pale fringes to feathers of upperparts.

Identification With long wings, forked tail, short legs, short, broad bill, and swallowlike flight, pratincoles rather unlike typical shorebirds, but still run along shore like shorebirds. Should be unmistakable in any plumage while on beach. Three very similar pratincoles breed in Eurasia and winter in Africa, Asia, and Australia, a so-far unrecorded one (Black-winged Pratincole, *Glareola nordmanni*) not an impossible vagrant to North America. Black-winged like Collared but has darker back, black instead of rufous underwings, no white edge on rear of wings, and shorter tail. Collared only one of three in which tail streamers often extend beyond wingtips. See Oriental Pratincole.

In Flight Wings dark with inner part of underwings reddish brown, secondaries white tipped; prominent white rump and rather long, mostly black forked tail, not unlike Barn Swallow. More likely to be mistaken for small, dark tern such as Black Tern than for any other shorebird. Much longer wings and more languid flight than swallows.

Voice Flight call a harsh, ternlike *kik kik kik* or *kitik*.

Behavior Runs along beaches like other shorebirds but spends much time hawking flying insects in air, often in flocks where common. Feeds on beach like plover with runs and stops; insects common prey.

Habitat Open plains, usually near water, in breeding season. Beaches, estuaries, and sometimes upland areas in nonbreeding season.

Range Breeds from Spain in scattered populations to central Asia. Winters in sub-Saharan Africa. Accidental in North America; one spent a winter in Barbados.

93.1 Breeding Collared Pratincole. Unusual shorebird with long wings and long, forked tail reaching wingtips. Bill relatively short and wide, aspect swallowlike or ternlike. Coloration distinctive, plain brown above with pale buff, black-bordered throat. Spain, Jun 1995 (Mike Lane).

93.2 Breeding Collared Pratincole. Long-winged, ternlike shorebird, distinguished by chestnut underwings, white line on rear edge of inner wing, and long, forked tail. Cyprus, Aug 2002 (Mike Danzenbaker).

94 Oriental Pratincole

(*Glareola maldivarum*)

Shorter tailed than its European counterpart, this species is no less swallowlike in aspect as it scours the sky for flying insects. Even when feeding on the beach like a standard shorebird, its looks are distinctive. Instead of lost southbound migrants, North American visitors have been northbound birds that overshot their normal breeding range.

Size Weight 85 g, length 23 cm (9 in), bill 1.4 cm, tarsus 3.5 cm.

Plumages Bill black, red at extreme base, legs dark olive. *Breeding adult* medium brown above, with rich brown breast, buffy sides, and white belly, with conspicuous buff throat patch bordered by black line. *Nonbreeding adult* with black line broken into fine streaks, bill base duller. *Juvenile* with bill base pale pink, whitish throat, mottled brown breast, pale fringes to feathers of upperparts.

Identification Differs from Collared Pratincole by less red at bill base (not reaching nostril), darker back, no white line on wing, substantially shorter tail (extends to wingtips or beyond in Collared, falls well short of wingtips in Oriental). Like Black-winged from above, with no white edge, but has chestnut underwings. Other differences include dull outer primary shaft (whitish in other species) and still shorter tail. Oriental slightly buffier below than other two in all plumages. Juveniles extremely similar.

In Flight All brown above with white rump, moderately forked tail, chestnut underwings. See Collared Pratincole.

Voice Like Collared Pratincole but usually two noted and may be lower pitched.

Behavior Like Collared Pratincole.

Habitat Like Collared Pratincole.

Range Breeds in central and southern Asia. Winters from southern Asia to northern Australia. Accidental in North America, two records from Bering Sea islands in spring.

94.1 Breeding Oriental Pratincole. Much like Collared Pratincole but tail much shorter, falls well short of wingtips. Slightly richer buffy brown. Tsushima, Nagasaki Prefecture, Japan, May 1995 (Tadao Shimba).

94.2 Nonbreeding Oriental Pratincole. Like breeding adult but bill without red, throat patch streaked, and line around it broken. Nishio, Aichi Prefecture, Japan, Sep 2003 (Tadao Shimba).

94.3 Juvenile Oriental Pratincole. Fine pale fringes on most feathers of upperparts, no black line around paler throat. Collared Pratincole identical, very unlikely for the two to occur in same parts of North America. Kyoto Prefecture, Japan, Aug 2002 (Tomokazu Tanigawa).

94.4 Adult Oriental Pratincole. Long-winged, almost ternlike shorebird with conspicuous white rump, shallowly forked tail. Much time spent feeding in air. Beidahe, China, May 1991 (Göran Ekström).

94.5 Adult Oriental Pratincole. Chestnut underwings as in Collared Pratincole but no white line on rear edge; tail not notably long. Note wing molt on upper-left bird. Nishio, Aichi Prefecture, Japan, Sep 2003 (Tadao Shimba).

References

These references provide more detail on the identification of shorebirds and their status in North America, as well as many more illustrations. Many earlier references to identification that are included in *Shorebirds of the Pacific Northwest* are not repeated here.

Abbott, S., S. N. G. Howell, and P. Pyle. 2001. First North American record of Greater Sandplover. *North American Birds* 55: 252–257.

American Birding Association. 2002. ABA Checklist: Birds of the Continental United States and Canada. American Birding Association, Colorado Springs, Colo.

American Ornithologists' Union. 1998. *Check–list of North American Birds*. 7th ed. American Ornithologists' Union, Washington, D.C.

Aversa, T. 2001. Nominate Rock Sandpiper at Ocean Shores, Washington. *North American Birds* 55: 242–244.

Baicich, P. J., and C. J. O. Harrison. 1997. *A Guide to the Nests, Eggs, and Nestlings of North American Birds*. Academic Press, London.

Benner, W. L. 1998. Broad-billed Sandpiper: Jamaica Bay, New York. *National Audubon Society Field Notes* 52: 513–516.

Bland, B. 1999. The Wilson's Snipe on Scilly revisited. *Birding World* 12: 56–61.

Brown, S., C. Hickey, B. Harrington, and R. Gill, eds. 2001. *United States Shorebird Conservation Plan*. Manomet Center for Conservation Sciences, Manomet, Mass.

Byrkjedal, I., and D. B. A. Thompson. 1998. *Tundra Plovers*. T & AD Poyser, London.

Carey, G., and U. Olsson. 1995. Field identification of Common, Wilson's, Pintail and Swinhoe's Snipes. *Birding World* 8: 179–190.

Chandler, R. J. 1989. *The Facts on File Field Guide to North Atlantic Shorebirds*. Facts on File, New York.

Cleeves, T. 1998. The Slender-billed Curlew in Northumberland—a new British bird. *Birding World* 11: 181–191.

Clements, J. F. 2000. *Birds of the World: A Checklist*. Ibis Publishing, Vista, Calif.

Colston, P., and P. Burton. 1988. *A Field Guide to the Waders of Britain and Europe, with North Africa and the Middle East*. Hodder & Stoughton, London.

Cramp, S., and K. E. L. Simmons, eds. 1983. *The Birds of the Western Palearctic*. Vol. 3. Oxford University Press, Oxford.

Daniels, G. G. 1978. Eurasian Curlew (*Numenius arquata*) on Martha's Vineyard, Massachusetts. *American Birds* 32: 310–312.

Dunn, J. L. 1993. The identification of Semipalmated and Common Ringed Plovers in alternate plumage. *Birding* 25: 238–243.

Engelmoer, M., and C. S. Roselaar. 1998. *Geographical Variation in Waders*. Kluwer Academic Publishers, Dordrecht, The Netherlands.

Faanes, C. A., and S. E. Senner. 1991. Status and conservation of the Eskimo Curlew. *American Birds* 45: 237–239.

Farrand, J., Jr. 1983. *The Audubon Society Master Guide to Birding*. Vol. 1. Alfred A. Knopf, New York.

Gibson, D. D., and B. Kessel. 1989. Geographic variation in the Marbled Godwit and description of an Alaska subspecies. *Condor* 91: 436–443.

Hayman, P., J. Marchant, and T. Prater. 1986. *Shorebirds: An Identification Guide to the Waders of the World*. Houghton Mifflin, Boston, Mass.

Heindl, M. T. 1999. The status of vagrant Whimbrels in the United States and Canada with notes on identification. *North American Birds* 53: 232–236.

Hirschfield, E., C. S. Roselaar, and H. Shirihai. 2000. Identification, taxonomy and distribution of Greater and Lesser Sand Plovers. *British Birds* 93: 162–189.

Hirst, P., and B. Proctor. 1995. Identification of Wandering and Gray-tailed Tattlers. *Birding World* 8: 91–97.

Jaramillo, A., and B. Henshaw. 1995. Identification of breeding-plumaged Long- and Short-billed dowitchers. *Birding World* 8: 221–228.

Jaramillo, A., R. Pittaway, and P. Burke. 1991. The identification and migration of breeding-plumaged dowitchers in southern Ontario. *Birders Journal* 1(1): 8–25.

Jehl, J. R., Jr. 1985. Hybridization and evolution of oystercatchers on the Pacific coast of Baja California. In *Neotropical Ornithology*, ed. P. A. Buckley, M. S. Foster, E. S. Morton, R. S. Ridgely, and F. G. Buckley. Ornith. Monogr. no. 36: 484–504.

Jones, C., and M. Holder. 1996. Bar-tailed and Black-tailed Godwits in Canada. *Birders Journal* 5: 184–193.

Kaufman, K. 1990. *A Field Guide to Advanced Birding*. Houghton Mifflin, Boston, Mass.

Kaufman, K. 1990. Curlew Sandpiper and its I.D. contenders. *American Birds* 44: 189–192.

Knopf, F. L. 1997. A closer look: Mountain Plover. *Birding* 29: 38–44.

Lakin, I., and K. Rylands. 1997. The Semipalmated Plover in Devon: the second British record. *Birding World* 10: 212–216.

Lanctot, R. B. 1995. A closer look: Buff-breasted Sandpiper. *Birding* 27: 384–390.

Lane, B. A. 1987. *Shorebirds in Australia*. Nelson Publishers, Melbourne.

Leader, P. 1999. Identification forum: Common Snipe and Wilson's Snipe. *Birding World* 12: 371–374.

Lethaby, N., and J. Gilligan. 1992. Great Knot in Oregon. *American Birds* 46: 46–47.

Marchant, S., and P. J. Higgins, eds. 1993. *Handbook of Australian, New Zealand, and Antarctic Birds*. Vol. 2. Oxford University Press, Melbourne.

McLaren, I. A., and B. Maybank. 1992. Apparent Broad-billed Sandpiper in Nova Scotia. *American Birds* 46: 48–50.

Mlodinow, S. 1993. Finding the Pacific Golden-Plover (*Pluvialis fulva*) in North America. *Birding* 25: 322–329.

Mlodinow, S. G. 1999. Spotted Redshank and Common Greenshank in North America. *North American Birds* 53: 124–130.

Mlodinow, S. G. 2001. Sharp-tailed Sandpiper. *Birding* 33: 330–341.

Mlodinow, S. G., S. Feldstein, and B. Tweit. 1999. The Bristle-thighed Curlew landfall of 1998: climatic factors and notes on identification. *Western Birds* 30: 133–155.

Mlodinow, S. G., and M. O'Brien. 1996. *America's 100 Most Wanted Birds*. Falcon Press, Helena, Mont.

Monroe, B. L., Jr., and C. G. Sibley. 1993. *A World Checklist of Birds*. Yale University Press, New Haven, Conn.

Morris, A. 1996. *Beautiful Beachcombers: Shorebirds*. NorthWord Press, Minnetonka, Minn.

Morrison, R. I. G., R. E. Gill, B. A. Harrington, S. Skagen, G. W. Page, C. L. Gratto-Trevor, and S. M. Haig. 2001. Estimates of shorebird populations in North America. Canadian Wildlife Service Occasional Paper no. 104. Environment Canada, Ottawa.

Mullarney, K., L. Svensson, D. Zetterström, and P. J. Grant. 1999. *Birds of Europe*. Princeton University Press, Princeton, N.J.

National Geographic Society. 2002. *Field Guide to the Birds of North America*. 4th ed. National Geographic Society, Washington, D.C.

Patterson, M. 1998. The great curlew fallout of 1998. *National Audubon Society Field Notes* 52: 150–155.

Paulson, D. 1993. *Shorebirds of the Pacific Northwest*. University of Washington Press, Seattle.

Payton, P. W. C. 2001. A closer look: Snowy Plover. *Birding* 31: 238–244.

Pitelka, F. A. 1950. Geographic variation and the species problem in the shorebird genus *Limnodromus*. *Univ. Calif. Publ. Zool.* 50: 1–108.

Prater, A. J., J. H. Marchant, and J. Vuorinen. 1977. *Guide to the Identification and Ageing of Holarctic Waders*. BTO Guide 17. British Trust for Ornithology, Tring.

Pringle, J. D. 1987. *The Shorebirds of Australia*. Angus & Robertson, North Ryde, Australia.

Rosair, D., and D. Cottridge. 1995. *Photographic Guide to the Shorebirds of the World*. Facts on File, New York.

Senner, S. E. 1998. A closer look: Surfbird. *Birding* 30: 306–312.

Sibley, D. A. 2000. *The Sibley Guide to Birds*. Alfred A. Knopf, New York.

Steel, J., and D. Vangeluwe. 2002. The Slender-billed Curlew at Druridge Bay, Northumberland, in 1998. *British Birds* 95: 279–299.

Stokes, D., and L. Stokes. 2001. *Beginner's Guide to Shorebirds*. Little, Brown & Company, Boston, Mass.

Szantyr, M. S. 1997. Semipalmated Sandpiper or Little Stint? A matter of degrees. *Birding* 29: 132–134.

Tuck, L. M. 1972. The snipes: a study of the genus *Capella*. Canadian Wildlife Service Monograph Series no. 5: 1–428.

Yovanovich, G. D. L. 1995. Collared Plover in Uvalde, Texas. *Birding* 27: 102–104.

Index